Building Maintenance FORMS, CHECKLISTS & PROCEDURES

Roger W. Liska and
Judith Morrison Liska

Prentice Hall

Library of Congress Cataloging-in-Publication Data

Liska, Roger W.
Building Maintenance Forms, Checklists and Procedures

Includes index.
1. Buildings—Maintenance—Forms. 2. Plant maintenance—Forms. 3. Buildings—Maintenance—Handbooks, manuals, etc. 4. Plant maintenance—Handbooks, manuals, etc. I. Liska, Judith Morrison II. Title
TH3351.L574 1988 658.2'02 88–22415
ISBN 0–13–093578–6

"This publication is designed to provide accurate and authoritative information in regard to the subject matter covered. It is sold with the understanding that the publisher is not engaged in rendering legal, accounting or other professional service. If legal advice or other expert assistance is required, the services of a competent professional person should be sought.

... From the Declaration of Principles jointly adopted by a Committee of the American Bar Association and a Committee of Publishers and Associations."

10 9 8 7 6 5 4 3 2

ISBN 0-13-093578-6

PRENTICE HALL
Paramus, NJ 07652

http://www.phdirect.com

About the Authors

Roger W. Liska is Chairman of the Department of Building Science at Clemson University, Clemson, South Carolina. He has more than 20 years' experience in the fields of engineering, construction, and maintenance of industrial, plant, and commercial buildings. His experience combines both teaching and administration at the university and junior college levels as well as the development and teaching of many seminars and workshops for industry, such as maintenance seminars for the Bell Telephone system.

Mr. Liska is a professional registered engineer. He is named in *Who's Who in Finance and Industry* and is a board member of the American Institute of Constructors and of the American Council for Construction Education. He is the author of the *Building and Plant Maintenance Deskbook* (Prentice Hall).

Judith Morrison Liska is a graphic supervisor at Clemson University, Clemson, South Carolina. She has over 13 years' experience in the design and production of commercial art. In addition, she has managed the development of major graphic art for large private and public organizations. Ms. Morrison Liska is active in graphic art associations and has won various awards for her work.

Dedication

This book is dedicated to those special people in the authors' lives who made it all possible—*our parents.*

A special thanks to Olivia Lane for her patience in the development of this book, to all those at the many companies who shared their ideas, and to Christina Burghard for her excellent editorial assistance.

What This Book Will Do for You

The *Handbook of Building and Plant Maintenance Forms and Checklists* has been carefully constructed to meet the needs of managers—including maintenance supervisors, plant managers, and vice-presidents of operations—responsible for overseeing the day-to-day performance of the maintenance function in a company. It will not only aid those managers who have a need for a workable system of documentation but also those who would like to design their own system. In addition, this *Handbook* can be useful for organizations that already have established documentation requirements as a means of supplementing such directives.

Lengthy research went into the compilation and design of the forms contained in this *Handbook*. And hundreds of companies of all sizes and types have shared their ideas, needs, and the actual maintenance forms in use in their companies today.

How to Use This Handbook

The *Handbook of Building and Plant Maintenance Forms and Checklists* has been organized in an easy-to-use format. Related maintenance forms and their applications have been grouped together into seven major sections:

1. Structural Systems
2. Interior and Exterior Building Finishes
3. Grounds and Landscaping Facilities
4. Housekeeping
5. Mechanical Systems and Equipment
6. Electrical Equipment and Systems
7. Maintenance Management

Within each major section there are a number of forms relating to the various activities that could be performed, such as inspection, lubrication, and troubleshooting. If you do not have forms to cover one or more of the activities, the specific format covered in this *Handbook* could be adapted for company use. If, however, you already have forms covering the activities in your company, the formats given here can be useful in revising them to be more encompassing and effective.

Twenty Ways This Handbook Will Help You

This *Handbook* provides you with a series of quality forms and checklists that can be used in the day-to-day management of the maintenance functions in any size organization to help make it cost effective. In addition, it:

1. Provides you with over 100 time-saving inspection checklists to document the current condition of materials, plant, and/or equipment. If material deterioration occurs at a later date, the information in the checklists can be very useful in reducing the time it takes to locate the cause, and thus the entire repair procedure will be less costly.

2. Helps utilize your maintenance budget more cost effectively through the use of tables. With these tables you can select the correct maintenance and/or repair procedure for the specific type of deterioration and thus eliminates "guesswork or trial-by-error" procedures.

3. Presents over 50 lists of required tools, material, and equipment for day-to-day housekeeping maintenance. Using these lists, the building manager can more quickly locate if any material deterioration is caused from the use of incorrect cleaning materials. In addition, they can be used to develop contract specifications for out-of-house housekeeping services. Without the availability of these lists, the maintenance manager would have to spend many hours "reinventing the wheel."

4. Contains dozens of checklists that can be used when planning a new building or altering an existing one. These checklists help to eliminate unnecessary and costly building items during new construction. In addition to this, the cost of maintaining the item(s) will also be eliminated.

5. Provides recommended intervals between specific maintenance operations and forms to be used in developing one's own man-hour allocation records.

6. Includes numerous charts and tables for the selection of finish building materials, such as floor and wall coverings for specific exposure and environmental conditions.

7. Helps solve repair problems faster through the use of the many tables covering the causes of concrete, steel, and timber deterioration—along with specific methods of diagnosing the causes. Hundreds of hours of investigative and decision-making time can be saved by using these tables.

8. Spells out maintenance procedures for materials and equipment. This results in easier to understand directions for the maintenance worker and cuts down on the amount of downtime needed to explain what is to be done.

9. Slashes the amount of time needed for numerous maintenance jobs, thus freeing up time to get to special projects, including critical repair work.

10. Simplifies the process of estimating the required labor, equipment and tools needed to perform specific maintenance and repair operations.

11. Presents various tables and forms to plan and schedule maintenance and repair operations more effectively.

12. Presents tested formats for controlling, and thus minimizing, maintenance costs.

13. Provides various personnel, accounting, evaluation, and other types of management forms to save time in developing or revising forms for record keeping and control.

14. Provides checklists to be used in the inspection of mechanical and electrical systems.

15. Supplies tables of frequencies of maintenance items for mechanical and electrical systems.

16. Gives a comprehensive one-stop list of sources to help eliminate the "search and seek" time needed for specialized equipment and materials. Many of these sources will provide free information and consulting services.

17. Features fingertip data to aid in selecting the correct cleaning and preserving agents, such as waxes and sealers, for specific exposure conditions and material types in your building.

18. Presents over 75 forms to be used in the organizing and staffing of the maintenance organization.

19. Includes instant reference tables which outline what substances are harmful to specific types of materials.

20. Provides hundreds of formats to aid in increasing the service life of equipment and facilities and enables you to bite the bullet on one of the most overlooked areas of savings in maintenance today—quality control.

In conclusion, the forms presented in this *Handbook* will assist you in performing your job in a more cost effective—and thus controllable—manner. This in turn will result in minimizing the cost of building and equipment maintenance and increasing the lifetime of the facilities. You should also note that all the forms presented in this book can be easily adapted for computer use with the help of a word processing software program.

Contents

What This Book Will Do for You . . . vii

Section 1 *USING FORMS IN THE MANAGEMENT OF MAINTENANCE ACTIVITIES* . . . 1

Maintenance Management Flow Chart • 2

Section 2 *STRUCTURAL SYSTEMS* . . . 5

A. Inventory Forms . . . 6

2.1: Inventory of Structural Members • 7

B. Inspection Forms . . . 9

2.2: Reinforced Concrete Inspection Checklist • 10
2.3: Structural Steel Inspection Checklist • 17
2.4: Structural Wood Inspection Checklist • 18
2.5: Truss Inspection Checklist • 20
2.6: Antenna Inspection Checklist • 21
2.7: Bridge and Trestle Inspection Checklist • 23
2.8: Chimney and Stack Inspection Checklist • 26
2.9: Elevated Tank Inspection Checklist • 28
2.10: Steel Power Pole and Structure Inspection Checklist • 29
2.11: Wood Pole and Accessories Inspection Checklist • 30
2.12: Loading Ramp Inspection Checklist • 31
2:13: Overhead Crane Inspection Form • 32

Section 3 *INTERIOR AND EXTERIOR FINISHES* . . . 33

A. Inventory Forms . . . 34

3.1: Inventory Building Finishes • 35
3.2: Roof Installation Summary Specifications • 37
3.3: Roofing—Historical Record • 39
3.4: Built-up Roofs—Historical Record • 40
3.5: Asphalt Roll Roofing Roofs—Historical Record • 42
3.6: Cement Composition Roofs—Historical Record • 44
3.7: Metal Roofs—Historical Record • 46
3.8: Slate Roofs—Historical Record • 48
3.9: Tile Roofs—Historical Record • 50
3.10: Wood Shingle Roofs—Historical Record • 52
3.11: Asphalt Shingle Roofs—Historical Record • 54
3.12: Painting—Historical Record • 56
3.13: Building Paint Work Record • 57
3.14: Glass Replacement Data • 58

B. Inspection Forms . . . 59

3.15: Building Inspection Checklist (Interior and Exterior) • 60
3.16: Building Rating Profile Summary Sheet • 63
3.17: Survey of Building Exterior • 68
3.18: Survey of Building Interior • 70

3.19: Roof Inspection Checklist • 74
3.20: Buildings Inspection Checklist (Except Roofs and Trusses) • 76
3.21: Survey of Roof Condition • 78
3.22: Condition of Roof Survey Report • 82
3.23: Wood Shingle Roof Inspection Checklist • 83
3.24: Built-Up Roof Inspection Checklist • 84
3.25: Asphalt Shingle Roof Inspection Checklist • 86
3.26: Cement Composition Roof Inspection Checklist • 87
3.27: Metal Roof Inspection Checklist • 88
3.28: Slate Roof Inspection Checklist • 90
3.29: Tile Roof Inspection Checklist • 91
3.30: Asphalt Roll Roofing Roof Inspection Checklist • 92

Section 4 LANDSCAPING AND GROUNDS FACILITIES **93**

A. Inventory Forms **94**

4.1: Grasses, Trees, and Shrubbery Inventory • 95
4.2: Exterior Ground Facilities Inventory • 97
4.3: Asphalt Surface Record • 99

B. Inspection Forms **100**

4.4: Grounds Inspection Checklist • 101
4.5: Pavement Inspection Checklist • 104
4.6: Cathodic Protection System Inspection Checklist • 105
4.7: Retaining Wall Inspection Checklist • 106
4.8: Seawall and Breakwall Inspection Checklist • 107
4.9: Tunnel and Underground Structure Inspection Checklist • 108
4.10: Fence and Wall Inspection Checklist • 110
4.11: Pier, Wharf, Quaywall, and Bulkhead Inspection Checklist • 111
4.12: Railroad Trackage Inspection Checklist • 112
4.13: Railroad Crossing Signal Inspection Checklist • 114
4.14: Storm Drainage System Inspection Checklist • 115
4.15: Refuse and Garbage Disposal Inspection Checklist • 117

C. Special Forms **118**

4.16: Building and Grounds Work Order • 119
4.17: Building and Grounds Work Schedule • 121
4.18: Vehicle Maintenance Report • 122
4.19: Vehicle Inspection Report • 124
4.20: Preventive Maintenance Record for Vehicles • 126
4.21: Vehicle Maintenance Schedule • 128
4.22: Building Herbicide Log • 130

Section 5 HOUSEKEEPING **131**

A. Schedule, Work Assignment, and Report Forms **132**

5.1: Master Housekeeping Schedule • 133
5.2: Building Cleaning Schedule • 134
5.3: Custodial Area Assignment • 136
5.4: Facilities Project Work Report • 138

5.5: Daily Periodic Area Cleaning Report • 139
5.6: Daily Floor Maintenance Work Report • 140
5.7: Daily Window Washing Work Report • 141
5.8: Carpet and Upholstery Work Report • 142

B. Inspection Forms 143

5.9: Room Housekeeping Checklist • 144
5.10: Custodial Area Sanitation Report • 145
5.11: Custodial Inspection Report—Floor Maintenance • 147
5.12: Custodial Inspection Report—Window Maintenance • 149
5.13: Housekeeping Inspection Form • 151
5.14: Supervisor's Checklist and Report • 154
5.15: Evaluation of the Custodial and Maintenance Program for School Buildings • 156
5.16: Exterior Maintenance Quality Assurance Evaluation • 162
5.17: Interior Maintenance Quality Assurance Evaluation • 164
5.18: Shop Areas Maintenance Quality Assurance Evaluation • 166

C. Special Forms 169

5.19: Snow Removal Call-In Sheet • 170
5.20: Building Pesticide Log • 171
5.21: Building Insecticide Log • 172
5.22: Supervisor's Daily Personnel Checklist • 173

Section 6 MECHANICAL SYSTEMS AND EQUIPMENT 175

A. Inventory Forms 176

6.1: Equipment Record • 177
6.2: Equipment Data Card • 178
6.3: Pump Data Sheet • 180
6.4: Heat Exchanger Data Sheet • 181

B. Inspection Forms 182

6.5: Heating, Ventilation, and Air-Conditioning Drawing Checklist • 183
6.6: Environmental Control System Pretesting and Startup Checklist • 187
6.7: Equipment Survey • 190
6.8: Annual Heating and Ventilating Systems Inspection Checklist • 193
6.9: Annual Plumbing Inspection Checklist • 200
6.10: Annual Fire Protection Equipment Inspection Checklist • 203
6.11: Detailed Building Air-Conditioning System Inspection Checklist • 205
6.12: Piping System Inspection Checklist • 208
6.13: Water Heater (All Types) Inspection Checklist • 210
6.14: Gas Distribution System Inspection Checklist • 212
6.15: Building Gas Heating System Inspection Checklist • 213
6.16: Paint Spray Booth Inspection Checklist • 215
6.17: Septic Tank Inspection Checklist • 216
6.18: Earthquake Valve and Pit Inspection Checklist • 217

6.19: Unit Heater Inspection Checklist • 218
6.20: Steam Trap Inspection Checklist • 219
6.21: Cooling Tower Inspection Checklist • 220
6.22: Small Turbine Inspection Checklist • 221
6.23: Large Turbine Inspection Checklist • 222
6.24: Bakery Equipment Inspection Checklist • 224
6.25: Portable Water Pumping Plant Inspection Checklist • 225
6.26: Fuel Distribution Facility Inspection Checklist • 226
6.27: Fuel Receiving Facility Inspection Checklist • 228
6.28: Fuel Storage Facility Inspection Checklist • 229
6.29: Walk-In Freezer Inspection Checklist • 231
6.30: Sewage Collection and Disposal Systems Inspection Checklist • 232
6.31: Refrigeration Equipment Inspection Checklist • 234
6.32: Safety Shower Inspection Checklist • 235
6.33: Swimming Pool Equipment Inspection Checklist • 236
6.34: Swimming Pool Inspection Checklist • 237
6.35: Chlorinator and Hypochlorinator Inspection Checklist • 238
6.36: Chemical Feed Equipment for Water Supply Inspection Checklist • 239
6.37: Eyewash Fountain Inspection Checklist • 241
6.38: Fresh Water Storage Inspection Checklist • 242
6.39: Ventilating Fan and Exhaust System Inspection Checklist • 243
6.40: Air-Handling Unit Inspection Checklist • 244
6.41: Liquid Filter Inspection Checklist • 246
6.42: Laundry Equipment Inspection Checklist • 247
6.43: Incinerator Inspection Checklist • 248
6.44: Ice Maker Inspection Checklist • 249
6.45: Heating and Ventilating Unit Inspection Checklist • 250
6.46: Heater and Console Control Inspection Checklist • 251
6.47: Pump Vacuum Producer Inspection Checklist • 252
6.48: Dishwashing Equipment and Accessories Inspection Checklist • 253
6.49: Dehumidification Unit Inspection Checklist • 255
6.50: Air Filter Inspection Checklist • 256
6.51: Air-Cooled Condenser Inspection Checklist • 257
6.52: Air Conditioning—Window and Fan Coil Unit Inspection Checklist • 258
6.53: Air-Conditioning Unit Inspection Checklist • 259
6.54: Air Compressor Inspection Checklist • 260
6.55: Aeration Equipment Inspection Checklist • 261
6.56: Dust Control System Check Sheet • 262
6.57: Pneumatic Temperature Controls Maintenance Checklist • 264

C. Special Forms 266

6.58: Mechanical Downtime Report • 267
6.59: Mechanical Services Overtime Report • 268
6.60: Lubrication Schedule • 269
6.61: Oil Consumption Form • 271

Section 7 ELECTRICAL EQUIPMENT AND SYSTEMS 273

A. Inventory Forms 274

7.1: Electrical Equipment Inventory Record • 275
7.2: Equipment and Repair Record • 276
7.3: Motor Maintenance Record • 277
7.4: Motor, Generator, and Control Services Record • 279

B. Inspection Forms 281

7.5: Electrical System (Buildings) Inspection Checklist • 282
7.6: Lighting System Inspection Checklist • 283
7.7: Distribution Transformer Inspection Checklist • 285
7.8: Power Transformer Inspection Checklist • 288
7.9: Electrical Ground Inspection Checklist • 293
7.10: Motor Inspection Checklist • 294
7.11: Motor/Generator Inspection Checklist • 296
7.12: Motor/Pump Inspection Checklist • 297
7.13: Motor/Pump Assembly (Hydraulic) Inspection Checklist • 298
7.14: Elevator Inspection Checklist • 300
7.15: Panel Board Inspection Checklist • 302
7.16: Value and Manhole (Electrical) Inspection Checklist • 303
7.17: Switch Gear Inspection Checklist • 304
7.18: Telephone Line (Open Wire) Inspection Checklist • 305
7.19: Time Clock Inspection Checklist • 306
7.20: Disconnecting Switch Inspection Checklist • 307
7.21: Electronic Air Cleaner Inspection Checklist • 309
7.22: Battery Inspection Checklist • 310
7.23: Lightning Arrestor Inspection Checklist • 311
7.24: Emergency Generator Set (Diesel or Gasoline) Inspection Checklist • 312
7.25: Emergency Lighting Inspection Checklist • 313
7.26: Doors (Power-Operated) Inspection Checklist • 314
7.27: Electrical Pothead Inspection Checklist • 315
7.28: Electrical Power Plant Inspection Checklist • 316
7.29: Electrical System (Waterfront) Inspection Checklist • 317
7.30: Oil Circuit Breaker Inspection Checklist • 318
7.31: Navigation Light Inspection Checklist • 319
7.32: Fire Alarm Panel Inspection Checklist • 320
7.33: Fire Alarm Box Light Inspection Checklist • 321
7.34: Fire Alarm Box Inspection Checklist • 322
7.35: Food Warmer/Grill Inspection Checklist • 323
7.36: Fuser Disconnect Inspection Checklist • 324
7.37: Underground Cable Inspection Checklist • 325

C. Special Forms 326

7.38: Notice of Electrical Power Shutdown • 327
7.39: Motor Job Sheet • 328
7.40: Light Tube Replacement Form • 329
7.41: Light Service Requisition • 330
7.42: Elevator Call Checklist • 331

Section 8 MAINTENANCE MANAGEMENT 333

A. Inventory Forms 334

8.1: Maintenance Specification • 335
8.2: Inventory Record • 337
8.3: Equipment Maintenance Record • 338
8.4: Equipment History • 339

B. Maintenance Work Request Forms 340

8.5: Immediate Maintenance Request • 341
8.6: Repair or Service Order • 342
8.7: Maintenance Service Request • 343
8.8: Minor Maintenance Work Order • 344
8.9: Maintenance Work Order • 345
8.10: Maintenance Work Request • 347
8.11: Work Order • 349

C. Work Order Summary Forms 350

8.12: Maintenance Work Sheet • 351
8.13: Job Order Register • 352
8.14: Labor and Material Work Order Summary • 353
8.15: Daily Service Log • 354

D. Inspection Forms 355

8.16: Maintenance Survey • 356
8.17: Plant Engineering and Maintenance Survey Report • 357
8.18: Maintenance Report • 358
8.19: Preventive Maintenance Inspection • 359

E. Inspection Summary Forms 360

8.20: Monthly Report of Preventive Maintenance Inspections • 361
8.21: Inspection Record Sheet • 362

F. Planning Forms 363

8.22: Maintenance Department Planner • 364
8.23: Maintenance Planning Sheet • 365
8.24: Comprehensive Maintenance Planning Sheet • 367

G. Estimating Forms 369

8.25: Work Sheet • 370
8.26: Summary Sheet • 372
8.27: Maintenance Cost Data Sheet • 374
8.28: Recapitulation Sheet • 376
8.29: Estimate Breakdown • 377
8.30: Budget Control Sheet • 379
8.31: Equipment Maintenance Cost Record • 380

H. Scheduling Forms 382

8.32: Project Master Schedule • 383
8.33: Master Schedule • 385

8.34: Monthly Master Maintenance Log • 386
8.35: Weekly Maintenance Work Schedule • 387
8.36: Daily Schedule • 388
8.37: Office Facilities Scheduled Maintenance Routine • 389
8.38: Schedule Control Sheet • 390

I. Purchasing, Requisition, and Expediting Forms 391

8.39: Bill of Material • 392
8.40: Maintenance Material Requisition • 393
8.41: Material Order Form • 394
8.42: Purchase Order • 395
8.43: Purchase Order Log Sheet • 396
8.44: Facilities Organization Material List • 397
8.45: Requisition for Stores Items • 399
8.46: Maintenance Equipment and Materials Inventory Checklist • 400
8.47: Maintenance Tool Inventory Checklist • 401
8.48: Tool Sign-Out List • 402

J. Control and Evaluation Forms 403

8.49: Maintenance Backlog Listing • 404
8.50: Maintenance Work-Order Backlog and Performance Report • 405
8.51: Weekly Labor Backlog Report • 407
8.52: Overtime Requirement • 408
8.53: Job Description Form • 409
8.54: Supervisor's Personnel File • 410
8.55: Employee Time Report • 411
8.56: Attendance Record • 412
8.57: Daily Absentee Report—Maintenance • 414
8.58: Notice of Violation of Rules • 415
8.59: Employee's Disciplinary Action • 416
8.60: Daily Travel Diary and Memorandum • 417
8.61: Inter-Office Communication • 418
8.62: New Maintenance Employee Training and Evaluation Schedule • 419
8.63: Supervisor's Daily Checklist • 420
8.64: Work Appraisal Form • 421
8.65: Equipment Operator's Report • 426
8.66: Equipment Failure and Accident Report • 427
8.67: Breakdown Report • 428
8.68: Maintenance Downtime Report • 430
8.69: Maintenance Failure History Analysis • 431
8.70: Service/Maintenance Contract Authorization • 432
8.71: Weekly Summary Outside Contractor Report • 433

K. Building Energy Use Forms 434

8.72: Monthly Plant Energy Consumption • 435
8.73: Energy Conservation Project Evaluation Summary • 436
8.74: Energy-Saving Survey • 437
8.75: Energy Conservation Capital Projects • 438

Index 439

Section 1

USING FORMS IN THE MANAGEMENT OF MAINTENANCE ACTIVITIES

In the early years of the maintenance business, all that seemed necessary to assure quality was the assignment of a job to personnel who possessed many years of successful experience in the business. No detailed records were kept; in fact, often many decisions that should have been made by supervisors and/or their employers were made by subordinates.

With the adoption of formal "preventive maintenance" programs, the maintenance manager realized the importance of utilizing formal documentation to insure a cost-effective program.

The Maintenance Management Flow Chart (Figure 1.1) shows the various places that forms are used within the different levels of any maintenance organization or system. The following six steps examine these levels.

STEP 1: INVENTORY FORMS

The beginning of a formal documentation process originates with an "Inventory Form" for every building component (doors, structural elements), distribution system (electrical, fresh water), and equipment and fixture (generators, motors, lights, sinks). It is important that a form exist on which such basic information is recorded as a description of the item, the date installed, manufacturer, who installed the item, and anything else that may have an effect on the inspection, maintenance, and repair of the specific item. The form should include space to record dates of inspection and performance of noted maintenance activities, such as cleaning and lubricating. Examples of these forms can be found throughout the book. See, for example, pages 7 and 35. The format of these forms are common in content for any building component, finish, system, equipment, and the like.

MAINTENANCE MANAGEMENT FLOW CHART

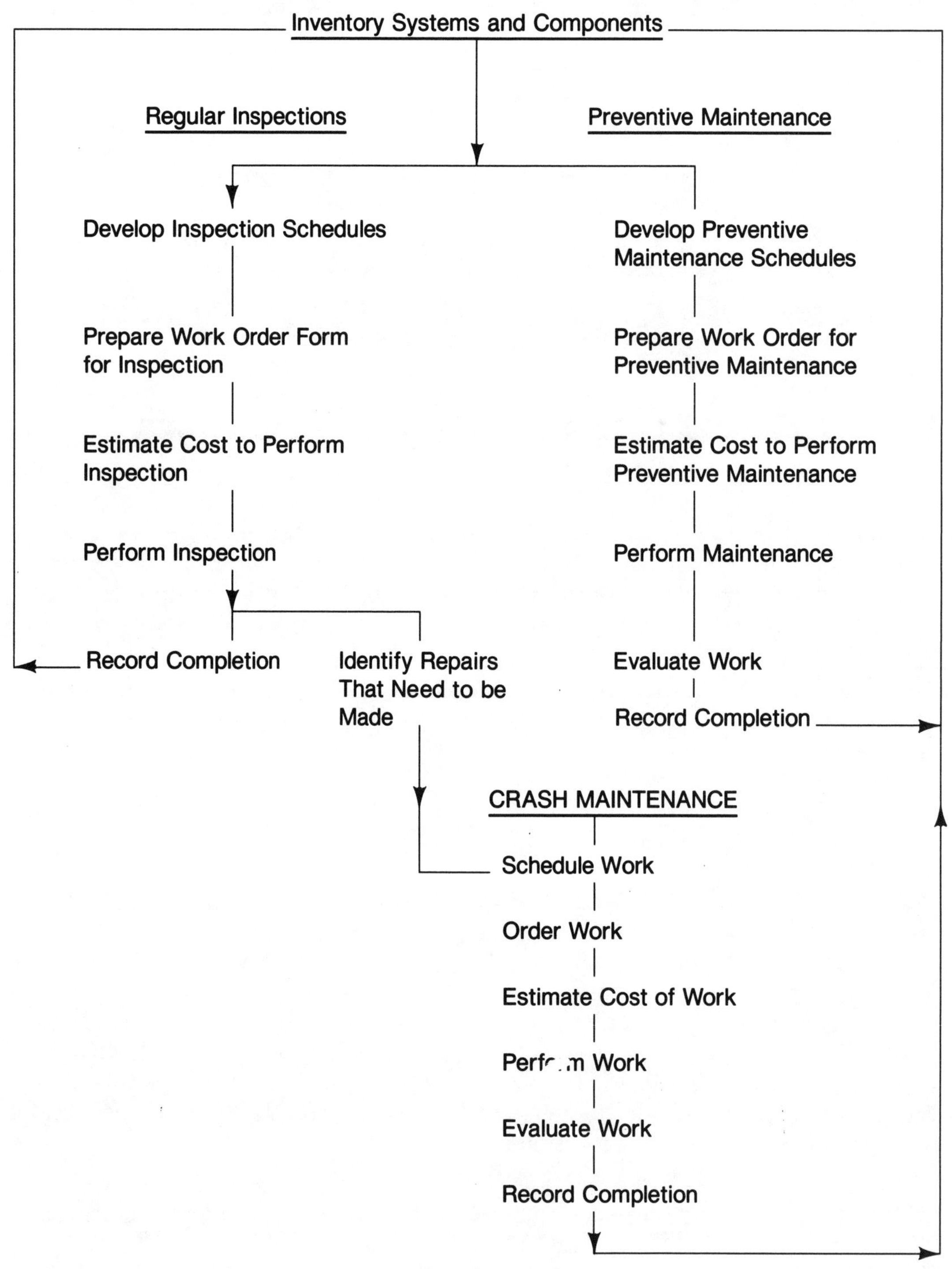

Figure 1.1

STEP 2: SCHEDULES

The next level in the flow chart is the preparation of inspection and routine preventive maintenance schedules. Both short- and long-term schedules need to be developed. Examples of these schedule formats can be found in Section 8.

STEP 3: WORK ORDERS

Whether inspection, planned, or crash maintenance, the request (order) to perform the activity must be documented and communicated to the necessary parties. This is done on a "work order" form. This form can be very simple or very complex in terms of the amount of information to be included on it. Work order forms can be found in Section 8.

STEP 4: PLANNING AND ESTIMATING FORMATS

Once the work that needs to be accomplished has been identified, the next step is to estimate what it will require in terms of time, personnel, equipment, tools, and material. This is done using formal estimating formats such as those found in Section 8 or may be placed on the work order form. For regularly scheduled activities, such as inspection or preventative maintenance, the estimate might be prepared at the time that the item is scheduled. In the case of nonroutine maintenance, the activity may have to be planned out systematically and a work description prepared. No matter what the activity is, the same forms can be used to plan and estimate the needed maintenance.

STEP 5: PERFORMING THE WORK

After receiving the final approval, the work is done. It is important that records be kept during the maintenance or repair activity as to time and resources actually used to accomplish the task. Here again, common formats are presented in Section 8 which are appropriate for any item.

STEP 6: EVALUATION OF COMPLETED WORK

Once the work is complete, it should be inspected to evaluate whether or not it was done in accordance to the designated specification or standard (work description). In addition, summaries need to be kept as to actual costs versus budgeted costs. General and special forms are presented in Section 8 to be utilized at this level of maintenance.

This brings us back to the beginning of the flow chart. Once the inspection or maintenance activity is complete, it should be recorded on the inventory form for the specific item. In addition to the steps listed here, there are related activities which can be documented such as employee training, employee evaluation, and the like.

Section 2

STRUCTURAL SYSTEMS

The inspection and maintenance of a building's structural system is critical to the safety of the occupants. Since most structural systems are very complex, the inventory activity must be formally documented. Every component of the structure including major and minor elements and connections need to be included in the inventory. This section provides formats that can be used to inventory components and record all inspections and repairs performed on the structural system.

The inspection of a typical structural system can be time consuming and costly. The reasons for this include the complexity of the system (not being able, in many cases, to actually see the structural elements due to finish materials covering or hiding them) and oftentimes the need to interrupt ongoing activities within the building. It is thus incumbent upon the inspector to perform the inspection in a well-planned manner so as to minimize the direct and indirect costs of the activity. This requires him or her to document all observations formally. Forms are included in this section that can be used to perform the inspection process.

The use of both the inventory and inspection forms will result in establishing efficient and effective maintenance control. Furthermore, if and when problems arise with any part of the structural system, the maintenance manager can refer to these documents for assistance in making any needed repairs. In essence, the use of the information on the forms will reduce the decision-making time of those involved in making any repairs. Finally, the data contained on the forms will be referred to prior to performing any renovations to all or part of the structural system. Design calculations will be performed in a more efficient manner knowing the kind of information shown on the inventory and inspection forms.

(See Section 8, *Maintenance Management,* for forms that can be used for planning, estimating, scheduling, and ordering and evaluating inspection, maintenance, and repair activities on structural components or systems.)

A. INVENTORY FORMS

The first form, 2.1, Inventory Structural Members, is to be used for inventorying all structural members. There should be one form completed for each member. An alternative to this method is to fill out one form for the structural system. This is recommended especially if the structural system is complex and the annual inspection is performed on the whole structure. As inspections, repairs, and other maintenance activities are performed on the member or system, the date accomplished along with other pertinent facts, such as repairs made, should be recorded on the inventory form. The person who will complete the form will depend on the type and size of the company. For example, in a relatively small company the inventory form may be completed by a plant engineer or his or her designated representative. While in a large company there most likely will exist a person in the Plant Engineering or Maintenance Department who will be responsible for completing and maintaining the forms.

This form is open-ended in that information will continue to be added to it. It serves as the historical basis for all construction and maintenance aspects of any one structural component or system within the building. The general information (such as Building Name, Type of Structure, and Type of Material) which must be recorded on the form can be obtained from "As-Built Drawings and Specifications." The other data to be recorded on the form is done so before and after any inspections and/or maintenance.

2.1: Inventory of Structural Members

INVENTORY OF STRUCTURAL MEMBERS

Building Name ______________________ Drawing Title* ______________

Drawing Numbers Which
Contain Structural Information ______________________

Date Constructed ______________________

Instructions: Briefly describe the type of structure including the material, frame type, type of connections, type of foundation and similar data.

Frequency of Inspection ______________________

Responsibility of Inspection ______________________

INSPECTION DATA

Inspection Performed By	Date	Comments

*Two sets of As-Built Drawings and Specifications should be on file.

2.1: *(Continued)*

MAINTENANCE DATA

Maintenance Performed By	Date	Describe Work Performed

OTHER COMMENTS:

B. INSPECTION FORMS

The next series of forms (2.2 through 2.13) are to be used in the inspection of various types of structures or building components. The person making the inspection should first record all the general types of information requested such as Date of Inspection, Item Being Inspected, and so on. The next step is the inspection and recording of observations on this document, along with any suggested follow-up activities that should take place such as an additional inspection, a need for immediate repair, or a recommendation for replacement. More than one day may be needed to complete the inspection process.

After the form is completed, it should be transmitted to the designated maintenance personnel. In addition, copies should be kept on file for future reference.

2.2: Reinforced Concrete Inspection Checklist

REINFORCED CONCRETE INSPECTION CHECKLIST

Item to Be Inspected ______________________________

Location ______________________________

Type ______________________________ Size ______________________________

Maintenance Inspector's Name ______________________________

Historical Data

Use of Item ______________________________

____

When Built ______________________________ Contractor ______________________________

Sketch Map

Instructions: Sketch the item being inspected, showing sunny and shady walls and well- and poorly-drained regions.

2.2: *(Continued)*

Instructions: Inspect the designated item(s) and make any notes relative to the maintenance and/or repair of the item. Take photographs where needed.

Present Condition of Structure	Date	Comments
Overall Alignment of Structure Settlement Deflection Expansion Contraction		
Portions Showing Distress (Beams, Columns, Pavement, Walls, etc., Subjected to Strains and Pressures)		
Surface Condition of Concrete General (Good, Satisfactory, Poor) Cracks Location and Frequency Type and Size Leaching, Stalactites Scaling Area, Depth Type Spalls and Popouts Number, Size, and Depth Type Extent of Corrosion or Chemical Attack Stains Exposed Steel Previous Patching or Other Repair		
Interior Condition of Concrete Strength of Cores Density of Cores Moisture Content (Degree of Saturation) Evidence of Alkali–Aggregate or Other Reaction Bond Aggregate, Reinforcing Steel, Joints Pulse Velocity Volume Change Air Content and Distribution		

2.2: *(Continued)*

Nature of Loading and Detrimental Elements	Date	Comments
Exposure Environment—Arid, Subtropical, Marine, Freshwater, Industrial, etc. Freezing and Thawing Wetting and Drying Drying under Dry Atmosphere Chemical Attack—Sulfates, Acids Abrasion, Erosion, Cavitation Electric Currents		
Drainage Flashing Weep Holes Contour		
Loading Dead Live Impact Vibration Traffic Index Other		
Soils (Foundation Condition) Stability Expansive Soil Settlement Restraint		
Original Condition of Structure	**Date**	**Comments**
Condition of Formed and Finished Surfaces Smoothness Air Pockets Sand Streaks Honeycomb Soft Area Early Structural Defects Cracking Plastic Shrinkage Settlement Cooling Curling Structural Settlement		

2.2: *(Continued)*

Materials of Construction	Date	Comments
Hydraulic Cement Type and Source Chemical Analysis (Obtain Certified Test Data if Available) Physical Properties		
Coarse Type, Source, and Mineral Composition (Representative Sample Available) Quality Characteristics Percentage of Deleterious Material Percentage of Potentially Reactive Materials Coatings, Texture, and Particle Shape Gradation, Soundness, Hardness Other Properties		
Fine Aggregate Type, Source, and Mineral Composition (Representative Sample Available) Quality Characteristics Percentage of Deleterious Material Percentage of Potentially Reactive Materials Coatings, Texture, and Particle Shape Gradation Other Properties		
Mixing Water Source and Quality		
Air-Entraining Agents Type and Source Composition Amount Manner of Introduction		
Admixtures Mineral Admixture Type and Source Physical Properties Chemical Properties Chemical Admixture Type and Source Physical Properties Chemical Properties		

2.2: *(Continued)*

Materials of Construction	Date	Comments
Admixtures (*Continued*) Chemical Admixture Type and Source Composition Amount		
Concrete Mixture Proportions Cement Content Proportions of Each Size Aggregate Water-Cement Ratio Water Content Chemical Admixture Mineral Admixture Air-Entraining Agent		
Properties of Fresh Concrete Slump Percent Air Workability Unit Weights Temperature		
Type Cast-In-Place Precast Prestressed		
Reinforcement Yield Strength Thickness of Cover Presence of Stirrups Use of Welding		

Construction Practices	Date	Comments
Storage and Processing of Materials Aggregates Grading Washing Storage Stockpiling Bins Cement and Admixtures Storage Handling		

2.2: *(Continued)*

Construction Practices	Date	Comments
Storage and Processing of Materials *(Continued)* Reinforcing Steel and Inserts Storage Placement		
Forming Type Bracing Coating Insulation		
Concreting Operation Batching Plant Type—Automatic, Manual, etc. Condition of Equipment Batching Sequence Mixing Type—Central Mix, Truck Mix, Job Mix, Shrink Mix, etc. Condition of Equipment Mixing Time Method of Transporting—Trucks, Buckets, Chutes, Pumps, etc. Equipment—Buckets, Elephant Trucks, Vibrators, etc. Weather Conditions—Time of Year, Rain, Snow, Dry Wind, Temperature, Humidity, etc. Site Conditions—Cut, Fill, Presence of Water, etc. Construction Joints Finishing Type—Slabs, Floors, Pavements, Appurtenances Method—Hand or Machine Equipment—Screeds, Floats, Trowels, Straightedge, Belt, etc. Additives, Hardeners, Water, Dust Coat, Coloring, etc. Curing Procedures Method—Water, Covering, Curing Compounds Duration Efficiency Form Removal (Time of Removal)		

2.2: *(Continued)*

Initial Physical Properties of Hardened Concrete	Date	Comments
Strength—Compressive, Flexural, Elastic Modulus Density Percentage and Distribution of Air Volume Change Potential Shrinkage or Contraction Expansion or Swelling Creep Thermal Properties Structural Section (Sketch and Thickness of Pavement Layers—Base, Subbase, etc.) Joints Type, Spacing, Design Condition Filling Material Faulting		
Structural Section (Sketch and Thickness of Pavement Layers—Base, Subbase, etc.) Joints Type, Spacing, Design Condition Filling Material Faulting Cracks Type (Longitudinal, Transverse, Corner) Size Frequency Patching Riding Quality Condition of Shoulders and Ditches		

2.3: Structural Steel Inspection Checklist

STRUCTURAL STEEL INSPECTION CHECKLIST

Building and Location ______________________________

Maintenance Inspector's Name ______________________________

Instructions: Inspect the noted structural items for the following:

- Cracks
- Localized distortion
- Paint or other protective coating failures
- Unusual wear, such as impact failure
- Misalignment
- Plumbness
- Corrosion
- Stress marks (show up in the cost of paint)
- Connections: loosening, weld cracks, corrosion
- General cleanliness

Item to Be Inspected	Date	Comments
Roof Trusses		
Open-Web Joists		
Bridging		
Bracing—Vertical and Horizontal		
Columns		
Beams and Girders		
Platform Supports		
Equipment Supports		
Stairs		
Bearing and Base Plates		
Siding Girts		
Roof Purlins		
Steel Decking		
Knee Braces		
Crane Rails		
Plate and Sheels		
Stacks		
Other		

2.4: Structural Wood Inspection Checklist

STRUCTURAL WOOD INSPECTION CHECKLIST

Building and Location ______________________________

Maintenance Inspector's Name ______________________________

Instructions: Inspect the noted structural items for the following:

- Decay
- Marine Borers
- Insect Damage
- Rodent Damage
- Fire Damage
- Splitting
- Checking
- Misalignment
- Soundness of Wood
- Loosened Connections
- Loosened Laminates

Item to Be Inspected	Date	Comments
Sills and Plates		
Beams and Girders (Ledgers)		
Rafters		
Joists		
Columns (Posts)		
Bridging		
Headers		
Studs		
Lintels		
Stairs		
Wood Trusses		
Nailers		
Connection Hardware		
Finish Wall and Ceiling Covering		
Finish Floor Covering		
Subfloors		
Fascia and Related Trim		

2.4 *(Continued)*

Item to Be Inspected	Date	Comments
Finish Trim		
Other Millwork		
Laminted Wood Members		
Other		

RECOMMENDATIONS:

2.5: Truss Inspection Checklist

TRUSS INSPECTION CHECKLIST

Maintenance Inspector's Name ____________________

Description and Location of Items Being Inspected ____________________

____

Instructions: Perform the specified inspection tasks. Insert the date when the activity is complete. Make any comments which are pertinent to future maintenance needs.

Item to Be Inspected	Date	Comments
From ground, check timber members for excessive number and size of knots; improper slope of grain; checks and splits in ends of web members; separation or slippage at joints; sag; signs of overloading; insect and fungus infestation.		
From truss, check timber members for loose bolts, split rings, shear plates, and fastening devices; checks and splits in bracing, chord members, splice plates, web members and filler blocks; loose tie rods; insect and fungus infestation.		
From ground, check steel members for twist, sag, and damage.		
From truss, check steel members for rupture, shearing or crushing of steel plates, members, bolts and rivets; loose bolts and rivets; broken welds.		
Check protective coatings for blistering, checking, cracking, scaling, flaking, mildew, bleeding, rust, and corrosion.		

Provide any suggestions or recommendations for follow-up activities or any other comments related to the inspection and maintenance activities.

2.6: Antenna Inspection Checklist

ANTENNA INSPECTION CHECKLIST

Maintenance Inspector's Name ______________________________

Description and Location of Items Being Inspected ______________________________

____

Instructions: Perform the specified inspection tasks. Insert the date when the activity is complete. Make any comments which are pertinent to future maintenance needs.

Item to Be Inspected	Date	Comments
Inspect foundations for cracked, broken, or spalled concrete; exposed reinforcing; movement or settlement; heaving from frost action.		
Check anchor bolts and straps for rust, corrosion, and loose, missing, or damaged parts.		
Check structural steel towers, ladders, and safety cages for rust, corrosion, and loose, missing, twisted, bowed, bent, or broken members.		
Check splices, bolts and rivets for rust, corrosion, loose, missing, other damage, broken welds.		
Inspect timber towers and ladders for loose, missing, twisted, bowed, cracked, split, rotted; termite or other insect infestation of wooden members.		
Inspect guys and anchorage for cracked, split, rotted; termite or other insect infestations, or looseness of wooden parts; metal parts for rust, corrosion, loose, missing, or other damage; frayed or broken strands and looseness of guys; adequate deadman anchorage.		
Check painted surface for rust, corrosion, cracking, scaling, peeling, wrinkling, alligatoring, chalking, fading, complete loss of paint.		
Check lights for proper operation.		
Inspect cleanliness of lights, shields, hoods, and receptacle fittings.		

2.6: *(Continued)*

Item to Be Inspected	Date	Comments
Check for sticking, binding, arcing, or burning of relay contacts; loose connections, missing parts of relays.		
Checking lightning rods and aerial terminals for damage from burning.		
Check conduits, terminals, and downleading cables for corrosion, loose or missing attachments to structures, other damage.		
Check bonding of aerial terminals, downleading cables, and ground connections.		
Check electrical continuity from aerial terminals through ground connections.		
Inspect for dirt, dust, grease, or other deposits on insulators; cracks, breaks, chips, or checking of the porcelain glaze.		
Check pulleys, winches, cables, ropes, or other elevating mechanisms and gear for proper operation.		

Provide any suggestions or recommendations for follow-up activities or any other comments related to the inspection and maintenance activities.

2.7: Bridge and Trestle Inspection Checklist

BRIDGE AND TRESTLE INSPECTION CHECKLIST

Maintenance Inspector's Name ______________________________

Description and Location of Items Being Inspected ______________________________

____

Instructions: Perform the specified inspection tasks. Insert the date when the activity is complete. Make any comments which are pertinent to future maintenance needs.

Item to Be Inspected	Date	Comments
Side slopes: failure to maintain slopes of $1^1/_2$ to 1 foot or more; soil erosion; inadequately protected with vegetation or mulch; concrete overlays (if applicable): cracking, spalling, broken areas, other damage.		
Bridge and foundation protective structures such as riprap, cribbing, bulkheads, dolphins, piles, insect and other pest infestation, decay, erosion, undermining, scouring, other damage.		
Drainage ditches: loose bottom and sides; improper side sloping; silting; failure to protect surrounding areas at outfalls from erosion.		
Roadway of approaches; cracked, broken, corrugated, and disintegrated concrete or bituminous surfaces; cracked, broken, or other damage to curb and gutter sections.		
Approach fill: settlement, particularly at joint between fill and structure.		
Fences, barricades, and railings at approaches: inadequacy or structural damage; missing or illegible load and speed limit signs.		
Drainage channels; erosion, scouring, accumulations of driftwood and debris above, below, and at structure; evidence of possible course diversion resulting from obstructions, erosion, or other.		

2.7: *(Continued)*

Item to Be Inspected	Date	Comments
Concrete foundation: cracks, scaling, disintegration, exposed reinforcing, wood piling and pads; missing, broken, ineffective bearing, decay, termite and other pest infestation; all foundations: scouring, undermining, settlement.		
Abutments and piers: cracks, breaks, scaling, spalling, disintegration, open joints, other damage; evidence of damage from impact and vibration; failure of expansion devices; damage from floating debris; ice and waterborne traffic.		
Timber framing: loose, missing, twisted, bowed, warped, split, checked, unsound members; deteriorated joints; rot, termite, and other insect infestation.		
Steel framing: rust, corrosion, loose, missing, bowed, bent, broken members.		
Concrete and masonry structures: weathering, cracks, spalling, exposed reinforcing; open, eroded, and sandy mortar joints; broken and missing stones.		
All superstructures: damage from floating debris, ice, and waterborne traffic; misalignment, both horizontal and vertical.		
Wood flooring: loose, missing, broken, rotten pieces; protruding nails and other fastenings; checkered wearing plates: loose, missing, or other damage.		
Structure roadways: cracked, broken, corrugated, disintegrated concrete, or bituminous surfaces.		
Concrete curbs and gutters and/or concrete or masonry handrails and handrail walls: loose, missing, and broken individual sections; misalignment; sandy and eroded mortar joints; loose or missing capstones; other damage.		
Expansion joints: improper sealing; loose or missing filler; failure to allow movement when filled with trash or debris.		

2.7: *(Continued)*

Item to Be Inspected	Date	Comments
Metal handrails: rust, corrosion, loose, missing, broken, misalignment, other damage.		
Bridge seats, bearing and cover plates: rust, corrosion, missing, loose, other damage.		
Rollers and other similar devices: rust, corrosion, inadequate lubrication; failure to allow movement.		
Cables: frayed, raveled, or broken strands; inadequate lubrication; defective anchorage; interference from overhanging objects.		
Splices, bolts, rivets, screws, and other connections: rust, corrosion, loose, missing, broken welds, other damage.		
Movable bridges: rust, corrosion, wear, inadequate lubrication. (Examine through complete operating cycle.)		
Utility supports: rust, corrosion, loose, missing, or broken parts.		
Utility lines: corrosion, leaks, sagging, insulation and waterproofing defects, mechanical damage.		
Painted surfaces: rust, corrosion, cracking, scaling, peeling, wrinkling, alligatoring, chalking, fading, complete loss of paint.		

Provide any suggestions or recommendations for follow-up activities or any other comments related to the inspection and maintenance activities.

2.8: Chimney and Stack Inspection Checklist

CHIMNEY AND STACK INSPECTION CHECKLIST

Maintenance Inspector's Name ______________________________

Description and Location of Items Being Inspected ______________________________

____

Instructions: Perform the specified inspection and maintenance tasks. Insert the date when the activity is complete. Make any comments which are pertinent to future maintenance needs.

Item to Be Inspected	Date	Comments
Inspect foundations for settlement and cracks.		
Inspect brick and concrete walls for weathering; cracking; deteriorated paint; damage from gases.		
Inspect caps for weathering; cracking; spalling; loose materials.		
Inspect exposed metal surfaces for rust, corrosion, and deteriorated paint; broken, loose, missing, damaged bolts, rivets, and welds.		
Inspect linings, supporting corbels, and baffles for cracks, spalling; damage.		
Inspect guys, anchorage, and bands for tautness; rust; corrosion; frayed, broken, loose, missing, damaged anchorage.		
Inspect ladders for rust; corrosion, paint scaling; anchorage; broken, loose, or missing ladder rungs.		
Inspect painters' trolley for wear, corrosion, damage to tiller ropes, pulleys, and pulley supports; pulley support anchorage.		
Inspect openings and cleanout doors, breechings and flues for cracking or spalling of the masonry surfaces; metal frames for distortion, rust, corrosion; broken, loose, missing, other damage to bolts, rivets, and welds.		
Inspect spark arrester screens for clogging with fly ash; rust, tears, and other damage; bolts and screws for rust, corrosion, loose, broken, or missing parts.		

2.8: *(Continued)*

Item to Be Inspected	Date	Comments
Inspect lights, hoods, reflectors, shields, and receptacle fittings for operation; repair or replace missing, loose, or damaged parts.		
Inspect conduit for breaks and other damage.		
Remove conduit inspection plates and examine internal connections for tightness and adequacy; relays for operation and for loose or weak contact springs; worn or pitted contacts; moisture.		

Provide any suggestions or recommendations for follow-up activities or any other comments related to the inspection and maintenance activities.

2.9: Elevated Tank Inspection Checklist

ELEVATED TANK INSPECTION CHECKLIST

Maintenance Inspector's Name ______________________________

Description and Location of Items Being Inspected ______________________________

____

Instructions: Perform the specified inspection and maintenance tasks. Insert the date when the activity is complete. Make any comments which are pertinent to future maintenance needs.

Item to Be Inspected	Date	Comments
Check receptacles, outlets, and conduits for protection against dirt, weather, and entrance of moisture; proper groundings.		
Check all threaded caps for security.		
Inspect wiring and electrical controls for loose connections; charred, broken, or wet insulation; evidence of short circuiting and other deficiencies. Tighten, repair or replace as required.		
Inspect recirculating pump for leakage and missing parts.		
Lubricate pump as required.		
Lubricate motor as required. *Do not over lubricate.*		
Check all covers and seals for missing bolts, screws, rust, corrosion, and other deficiencies.		
Check for proper identification and markings.		

Provide any suggestions or recommendations for follow-up activities or any other comments related to the inspection and maintenance activities.

2.10: Steel Power Pole and Structure Inspection Checklist

STEEL POWER POLE AND STRUCTURE INSPECTION CHECKLIST

Maintenance Inspector's Name ______________________________

Description and Location of Items Being Inspected ______________________________

____

Instructions: Perform the specified inspection tasks. Insert the date when the activity is complete. Make any comments which are pertinent to future maintenance needs.

Item to Be Inspected	Date	Comments
Check concrete bases, pads, and anchor bolts for cracks, breaks, settlement, ponding, rust and corrosion, loose or missing nuts and bolts.		
Check street light standard handholes and bell interiors for rust, corrosion, damage, and leakage.		
Check poles, structures, crossarms, and beams for rust and corrosion, misalignments, and loose bolts and pins.		
Check guys and anchors for rust and corrosion, damaged hardware, missing or damaged insulators, damaged or corroded guy shields, and excessive sag or tension of guys.		
Check ground wire for rust, corrosion, looseness, proper support and protection, and continuity.		

Provide any suggestions or recommendations for follow-up activities or any other comments related to the inspection and maintenance activities.

2.11: Wood Pole and Accessories Inspection Checklist

WOOD POLE AND ACCESSORIES INSPECTION CHECKLIST

Maintenance Inspector's Name ________________________________

Description and Location of Items Being Inspected ________________________________

____

Instructions: Perform the specified inspection and maintenance tasks. Insert the date when the activity is complete. Make any comments which are pertinent to future maintenance needs.

Item to Be Inspected	Date	Comments
Soundtest poles with hammer for hollowness or decay from ground line to the highest point reached from standing position.		
Check poles for decay, damage, splits, alignment, and insect and fungus infestation.		
Check crossarms and buckarms for splits, burns, decay, damage, and insect and fungus infestation.		
Check insulators and pins for cracks, breaks, looseness, rust, corrosion, and damage.		
Check tie wires and line wires for looseness, chafing, slippage, or other damage.		
Check ground wires for corrosion, frayed or broken strands, signs of overheating, and proper connection to ground rod.		
Check protective moldings for looseness, cracks, and damage.		
Check guy wires for looseness, corrosion, and damage.		

Provide any suggestions or recommendations for follow-up activities or any other comments related to the inspection and maintenance activities.

2.12: Loading Ramp Inspection Checklist

LOADING RAMP INSPECTION CHECKLIST

Maintenance Inspector's Name ______________________

Designation and Location of Item Inspected ______________________

____

Instructions: Perform the specified inspection and maintenance tasks. Insert the date when the activity is complete. Make any comments which are pertinent to future maintenance needs.

Item to Be Inspected	Date	Comments
Check main relief valve and adjust as necessary to maintain proper operating pressure. Record pressure prior to adjustment.		
Check relief valve for sticking.		
Check circulation by observing fluid in reservoir.		
Check temperature of oil for proper operation of heat exchanger and record.		
Check unit for noise and vibration.		
Check pump volume control for operation.		
Check oil level in reservoir and add oil if necessary.		
Check system for leaks.		
Check shaft couplings for wear and alignment.		
Check motor bearing for heating and/or noise.		
Check pipe hangers and supports. Tighten where necessary.		
Check motor starter and control contacts. Replace where necessary.		
Lubricate electric motor where applicable. *Do not over lubricate.*		
Check shaft alignment and condition of coupling. Align, repair, or replace if required.		
Inspect electrical equipment for condition, dust, and moisture.		

Provide any suggestions or recommendations for follow-up activities or any other comments related to the inspection and maintenance activities.

2.13: Overhead Crane Inspection Form

OVERHEAD CRANE INSPECTION FORM

Crane Designation ______________ Maintenance Inspector's Name ______________ Date of Inspection __________

Crane Location ______________

Parts	Indicate Condition*	Indicate Any Follow-up Action Needed
Drums, Chains, Cables, and Hooks		
Wheels and Flanges		
Brakes and Bells		
Sweep Brushes and Bumpers		
Track		
Draw Bars and Push Poles		
Electrical Equipment		
Foot Walks and Railings		
Warning Signs		
Condition of other parts not specified above		
Does operator consider crane safe?		
Should crane be shut down immediately until repaired?		
Track Clamps, Bridge, and Trolley		

***CONDITION CODE**

Good — Good condition; no repairs required.
Fair — Can operate safely. Adjustments or repairs can wait until next scheduled down time.
Poor — Repairs needed immediately. Take out of service. Continued use unsafe. Bigger damage risk.

Section 3

INTERIOR AND EXTERIOR FINISHES

The finish materials inside and outside a building serve many purposes. Some of the major ones provide an appealing visual environment, protect the structural components, serve as wearing surfaces, and provide enclosed space in which the occupants are protected from inclement weather. The cost of all finish materials needed to complete any building is relatively high as compared to the costs of the other components and systems found in the facility. Furthermore, studies have shown that the cost to maintain and repair finish materials over their lifetime will far exceed their initial in-place cost. The exact amount depends on the effectiveness of the maintenance program. If the level of maintenance is low to nonexistent, the materials will most likely have to be repaired or replaced prematurely.

To minimize this from occurring, it is important that a preventive maintenance program is in place. The basis of such a program must be a complete inventory of all the finish materials of which the facility is comprised. This inventory must be developed in a formal manner and be documented for future use.

In addition, an effective and efficient preventive maintenance program must include inspections performed on a regular frequency for all interior and exterior building finishes. The results of the inspections provide information to the maintenance management as to the most appropriate housekeeping and needed repair procedures that must be performed to insure that the design lifetime of the respective material is realized. The inspections must be performed in an organized manner utilizing standard formats.

This section provides forms useful in inventorying interior and exterior building finishes and in recording all inspections, maintenance, and repair activities on the specific finish. Also included are forms for use in the inspection process.

A. INVENTORY FORMS

To control the thousands of different finish materials that can be used on the interior and exterior of a building, it is essential to develop and maintain a complete inventory. Form 3.1, Inventory Building Finishes, presents a format for such a task. One form should be completed for each interior and exterior finish. For example, there should be one form for wood paneling and one for plastic-coated paneling. Even though this will result in a large number of forms, the benefits outweigh the disadvantages once maintenance problems begin.

As inspections, repairs, and other maintenance activities are performed on each item, the date accomplished, along with other pertinent facts such as maintenance activities performed, should be recorded on the inventory form. The person who will complete the form will depend on the type and size of the company. This form is open-ended in that information will continue to be added to it. It serves as the historical basis for all maintenance aspects of the interior and exterior building finishes.

There are many computer software programs available today that can help you to inventory building finishes along with other pertinent information such as dates of inspection, required follow-up maintenance activities, and when major repairs were performed.

Roof finishes have been identified as one of the most named maintenance problems by building owners. A comprehensive series of forms (3.2 through 3.11) have been included for the purpose of creating historical data on each type of roof. Forms that are to be used in the inspection of them follow in Section B.

The Roof Installation Summary Specifications (3.2) and all historical record forms (3.3 through 3.11) are used to document the composition of a designated roofing system. They are very similar to an inventory form. To complete these forms, first provide the general kinds of information requested in the appropriate spaces. Next, check the boxes that pertain to the appropriate component of construction of which the roof is comprised. Also, provide any other information requested. This should be done during the construction process. If the building is already complete, refer to the As-Built Drawings and Specifications to obtain the information. When roofing modifications are performed, be sure to update the respective forms.

Other inventory-type forms have been included for painting and glass replacement. They are completed in the same manner as for the other inventory forms. The Painting Historical Record and the Building Paint Work Record (3.12 and 3.13) are nothing more than records (or histories) of the painting program for the building(s). They are essentially an inventory of the type of paint used and date applied. Therefore, they should be completed and used as any other inventory form. The data contained in them will assist management in making more efficient decisions on when to repaint surfaces. Finally, the Glass Replacement Data form (3.14) allows the user to keep track of all glass replacement in the facility. Its use is similar to that of an inventory form.

3.1: Inventory Building Finishes

INVENTORY BUILDING FINISHES

Building Name ______________________ Finish ______________

Finish Appears on
Drawing No.* ______________ Date Installed ______________

Instructions: Briefly describe the location of the finish in the building and method of installation.

Frequency of Inspection ______________________

Responsibility of Inspection ______________________

INSPECTION DATA

Inspection Performed By	Date	Comments

*Two sets of As-Built Drawings and Specifications should be kept on file.

3.1: *(Continued)*

REPAIR (MAINTENANCE) DATA

Maintenance Performed By	Date	Comments

HOUSEKEEPING DATA

Instructions: Briefly describe the housekeeping activities to be performed to maintain the material along with frequence, cleaning products used, etc.

3.2: Roof Installation Summary Specifications

ROOF INSTALLATION SUMMARY SPECIFICATIONS

Facility ______________________ Date Constructed (Roof) ____________

Form Completed By ______________________ Date Completed ____________

ROOF SYSTEM

☐ Single Ply	☐ Ballasted	☐ Unballasted	
☐ IRMA			
☐ Built-Up			
Bitumen	☐ Asphalt	☐ Coal Tar	
Felts	☐ Fiberglass	☐ Organic	
Surface	☐ Gravel	☐ Slag	☐ Crushed Stone
	☐ Embedded Mineral	☐ Smooth	
Vapor Barrier	☐ Yes	☐ No	
Base Flashing	☐ Mineral Surface	☐ Elastic Sheeting	☐ Built-up
	☐ Other ____________		
Counter-Base Flashing		☐ Surface Mount	☐ Reglet
Slope	☐ Less than $\frac{1}{2}''$/Ft.	☐ More than $\frac{1}{2}''$	☐ Flat

Roof Area ____________ Sq. Ft.

Manufacturer & Spec. ______________________________

Date Installed ______________________

INFORMATION

Deck	☐ Steel	☐ In-Situ Concrete	☐ Precast Concrete
	☐ Poured	☐ Gypsum	☐ Wood
	☐ Other ______________________		Thickness ______
	Dead Load ____________	Live Load ____________	
Insulation	☐ Fiberglass	☐ Fiberboard	☐ Perlite
	☐ Other ______________________		Thickness ______
	Type Vents ______________________		
Equipment Suports		☐ Flashed Curbs	☐ Wood Blocking
	☐ Columns	☐ Other ____________	

3.2: *(Continued)*

Drainage System	☐ Roof Drains	Qty. ________	
	☐ Scuppers–Overflow Drains	Qty. ________	
	☐ Gutters–Downspouts		
Pipe–Conduit	☐ On Supports	☐ On Wood Blocking	
	☐ On Membrane		
Parapet Walls	Height ____________	☐ Brick	☐ Plaster
	☐ Other ______________________________		
Coping	☐ Metal	☐ Stone	☐ Tile
	☐ Concrete	☐ Other ____________________	
Are Ducts Present	☐ Yes	☐ No	☐ Insulated
	☐ Uninsulated		

COMMENTS:

3.3: Roofing—Historical Record

ROOFING—HISTORICAL RECORD

Prepared By ______________________________ Date __________

1. Facility ______________________________
2. Year Building Was Previously Roofed ______________________________
3. Date Roof (current) Was Applied ______________________________
4. Contract or Job Order No. ______________________________
5. Contractor's Name ______________________________
6. Type of Roof ______________________________
7. Area of Roof in Squares ______________________________
8. Cost per Square ______________________________
9. Brand Name of Roofing Material Used ______________________________
10. Type of Roof Sheathing ______________________________
11. Condition of Roof Sheathing ______________________________
12. Surface Preparation: Complete Tearoff __________ Reroof Over Old __________
    Other ______________________________
13. Quantities of Materials Used ______________________________
    ____
14. Flashing Type ____________________ Linear Feet __________
15. Insulation Type ____________________ Brand Name __________
16. Weather Conditions During Construction ______________________________
17. General Remarks:

FORM 3.3

Key Use of Form:

To record information about the roof construction which may be needed at a later date for use in repair and maintenance activities.

Who Prepares:

Plant engineering/maintenance personnel.

Who Uses:

Plant engineering.

How to Complete:

Provide the information requested in the spaces. Obtain the information from the roofing contractor, drawings, and technical specifications.

Alternative Forms:

Forms similar to this one, but revised to make it more useful for a specific company.

3.4: Built-up Roofs—Historical Record

BUILT-UP ROOFS—HISTORICAL RECORD

Form Completed By ______________________ Date ____________

Building ____________ Used for ______________________

☐ Permanent ☐ Temporary Year Roof Was Applied ________

TYPE OF ROOF DECK

☐ Wood ☐ Concrete Slab ☐ Concrete

☐ Gypsum Slab ☐ Gypsum Plank ☐ Steel

SLOPE OF ROOF

☐ Flat ☐ Sloped _____ In. per foot

AREA OF ROOF ____________ Squares

TYPE OF BUILT-UP ROOF

Asphalt ☐ Surfaced ☐ Unsurfaced ☐ Cold Process

☐ Wide Selvage

Coal Tar Pitch ☐ Yes ☐ No

KIND OF SURFACING

☐ Slag ☐ Gravel ☐ Crushed Stone

☐ Promenade Tile ☐ Slate Slabs

☐ Mineral Surfaced Cap Sheet

☐ Smooth Surfaced Cap Sheet

☐ Other______________________

NUMBER OF PLIES OF FELT

☐ 2 ☐ 3 ☐ 4

☐ 5 ☐ Other ________

KIND OF FELT

☐ Organic ☐ Coated ☐ Uncoated

☐ Asphalt ☐ Coated ☐ Uncoated

INSULATION

☐ Yes ☐ No

Type of Insulation ______________ Thickness ______________

Where Placed ______________________________

Vapor Seal ☐ Yes ☐ No Type ________

3.4: *(Continued)*

FLASHINGS

Base Flashing	☐ Metal	Kind ____________________
	☐ Composition	Kind ____________________
	☐ Other ____________________	
Counter or Cap Flashing	☐ Yes	☐ No
Through Wall Flashing	☐ Yes	☐ No
	☐ Metal	Kind ____________________
	☐ Composition	Kind ____________________
	☐ Other ____________________	
Flashing Block	☐ Yes	☐ No

PREVIOUS MAINTENANCE (Describe briefly with dates)

Roof Membrane:

Flashings:

PREVIOUS REPAIRS (Describe briefly with dates)

Roof Membrane:

Flashings:

GENERAL REMARKS:

3.5: Asphalt Roll Roofing Roofs—Historical Record

ASPHALT ROLL ROOFING ROOFS—HISTORICAL RECORD

Form Completed By ______________________ Date __________

Building ______________________ Used for ______________________

☐ Permanent ☐ Temporary Year Roof Was Applied __________

KIND OF ROOF DECK

☐ Sheathing Boards Thickness ______

☐ Plywood Thickness ______

UNDERLAYER

☐ None ☐ Saturated Felt ☐ Paper

☐ Asphalt Shingles ☐ Wood Shingles

☐ Other ______________________

SLOPE OF ROOF __________ In. Per Foot

AREA OF ROOF __________ Squares

TYPE OF ROLL ROOFING

☐ Mineral Surfaced ☐ Smooth Surfaced

☐ Wide Selvage

METHOD OF LAYING

☐ 2 In. Lap ☐ 4 In. Lap ☐ 19 In. Lap

☐ Concealed Nails ☐ Exposed Nails

TYPE OF LAP CEMENT ☐ Hot Applied ☐ Cold Applied

COLOR OF ROOFING GRANULES ______________

FLASHINGS

Valley Flashings ☐ Roll Roofing ☐ Asphalt Shingles

☐ Metal Kind of Metal ______________

Chimney Flashings ☐ Roll Roofing ☐ Metal

Kind of Metal ______________________

3.5: *(Continued)*

PREVIOUS MAINTENANCE (Describe briefly with dates)

Asphalt Roll Roofing:

Flashings:

PREVIOUS REPAIRS (Describe briefly with dates)

Asphalt Roll Roofing:

Flashings:

GENERAL REMARKS:

3.6: Cement Composition Roofs—Historical Record

CEMENT COMPOSITION ROOFS—HISTORICAL RECORD

Form Completed By ______________________________ Date ____________

Building ________________________ Used for ____________________________

☐ Permanent ☐ Temporary Year Roof Was Applied ________

TYPE OF ROOF ☐ Corrugated Sheets ☐ Shingles

TYPE OF SHINGLE

☐ Dutch Lap ☐ Hexagonal ☐ American

☐ Multiple Unit

TYPE OF DECK

Shingle ☐ Wood ☐ Other ______________________

Corrugated ☐ Wood ☐ Other ______________________

Purlin Spacing ________________ In.

SLOPE OF ROOF ________________ In. per foot

AREA OF ROOF ________________ Squares

TYPE OF UNDERLAYER

☐ Asphalt saturated felt

☐ Asphalt saturated organic felt

☐ Other ____

Weight of Felt ________ Pounds

KIND OF FASTENERS

Shingles ☐ Nails ☐ Other _________ Size _________

Corrugated Sheets ☐ Clips Type _________________________

☐ Screws Size _________________________

☐ Bolts Size _________________________

☐ Nails Size _________________________

☐ Other Size _________________________

3.6: *(Continued)*

FLASHINGS

Valley Flashings	☐ Metal	Kind ______________
	☐ Other ______________	
Vent Flashings	☐ Metal	Kind ______________
	☐ Other ______________	
Drip Edge	☐ Metal	Kind ______________
Chimney Flashings	☐ Metal	Kind ______________
	☐ Other ______________	

PREVIOUS MAINTENANCE (Describe briefly with dates)

Cement Composition Roofing:

Flashings:

PREVIOUS REPAIRS (Describe briefly with dates)

Cement Composition Roofing:

Flashings:

GENERAL REMARKS:

3.7: Metal Roofs—Historical Record

METAL ROOFS—HISTORICAL RECORD

Form Completed By ______________________ Date ____________

Building ____________________ Used for ____________________

☐ Permanent ☐ Temporary Year Roof Was Applied ________

TYPE OF DECK

☐ Wood ☐ Metal ☐ Other ________

☐ Solid ☐ Open ☐ Purlins

Purlin Spacing ________

SLOPE OF ROOF ________ In. per foot

AREA OF ROOF ________ Squares

KIND OF METAL

☐ Tin (Terne) Weight Coating ____________

☐ Copper Weight Per Sq. Ft. ____________

☐ Galvanized Steel Weight ____________

☐ Aluminum Gage ____________

☐ Protected Metal Kind ____________

TYPE OF METAL ROOF

☐ Flat Sheets ☐ Corrugated Sheets

☐ Special Shapes ☐ Shingles or Tiles

TYPES OF SEAMS

☐ Batten ☐ Standing ☐ Other ________

☐ Flat Soldered: ☐ Yes ☐ No

TYPE OF FASTENERS

☐ Nails ☐ Clips ☐ Screws

☐ Cleats ☐ Other ____________

FLASHINGS

Valley Flashings (Describe):

Vent Flashings (Describe):

Chimney Flashings (Describe):

3.7: *(Continued)*

PREVIOUS MAINTENANCE (Describe Briefly with Dates)

Metal Roofing:

Flashings:

PREVIOUS REPAIRS (Describe Briefly with Dates)

Metal Roofing:

Flashings:

GENERAL REMARKS:

3.8: Slate Roofs—Historical Record

SLATE ROOFS—HISTORICAL RECORD

Form Completed By ____________________ Date ____________________

Building ____________________ Used for ____________________

☐ Permanent ☐ Temporary Year Roof Was Applied __________

TYPE OF SLATE ____________________ Approx. Thickness ____________________

COLOR OF SLATE ____________________ Approx. Wt. Per Sq. ____________________

SLOPE OF ROOF ______________ In. Per Foot

ROOF DECK ☐ Wood Thickness ________ Width __________

☐ Other ____________________

UNDERLAYER

☐ Asphalt saturated felt Weight __________

☐ Asphalt saturated organic felt Weight __________

☐ Other ____________________

TYPE OF FASTENERS

☐ Nails Size __________

☐ Other ____________________

FLASHINGS

Valley Flashings ☐ Metal Kind ____________________

☐ Other ____________________

Vent Flashings ☐ Metal Kind ____________________

☐ Other ____________________

Drip Edge ☐ Metal Kind ____________________

Chimney Flashings ☐ Metal Kind ____________________

☐ Other ____________________

PREVIOUS MAINTENANCE (Describe Briefly with Dates)

Slate Roofing:

Flashings:

3.8: *(Continued)*

PREVIOUS REPAIRS (Describe Briefly with Dates)

Slate Roofing:

Flashings:

GENERAL REMARKS:

3.9: Tile Roofs—Historical Record

TILE ROOFS—HISTORICAL RECORD

Form Completed By ____________________ Date ____________________

Building ____________________ Used for ____________________

☐ Permanent ☐ Temporary Year Roof Was Applied __________

TYPE OF TILE

☐ Shingle ☐ Spanish ☐ Interlocking

☐ Other ____________________

Thickness __________ Wt. per square __________

Color ____________________

SLOPE OF ROOF __________ In. per foot

AREA OF ROOF __________ Squares

ROOF DECK

☐ Wood Thickness __________

Width of Sheathing Boards __________

☐ Other ____________________

UNDERLAYER

☐ Asphalt saturated felt Weight __________

☐ Asphalt saturated organic felt Weight __________

☐ Other ____________________

TYPE OF FASTENERS

☐ Nails Size __________

☐ Other ____________________

FLASHINGS

Valley Flashings ☐ Metal Kind __________

☐ Other ____________________

Vent Flashings ☐ Metal Kind __________

☐ Other ____________________

Drip Edge ☐ Metal Kind __________

Chimney Flashings ☐ Metal Kind __________

☐ Other ____________________

3.9: *(Continued)*

PREVIOUS MAINTENANCE (Describe Briefly with Dates)

Tile Roofing:

Flashings:

PREVIOUS REPAIRS (Describe Briefly with Dates)

Tile Roofing:

Flashings:

GENERAL REMARKS:

3.10: Wood Shingle Roofs—Historical Record

WOOD SHINGLE ROOFS—HISTORICAL RECORD

Form Completed By ______________________ Date ______________________

Building ______________________ Used for ______________________

☐ Permanent ☐ Temporary Year Roof Was Applied __________

KIND OF ROOF DECK
- ☐ Sheathing Boards — Thickness ______
- ☐ Single Lath — Width of Spcg. ______ Thickness ______
- ☐ Plywood — Thickness ______

UNDERLAYER
- ☐ None ☐ Saturated Felt ☐ Paper
- ☐ Wood Shingles ☐ Asphalt Roll Roofing
- ☐ Asphalt Shingles
- ☐ Other ______________________

SLOPE OF ROOF ______________ In. Per Foot

AREA OF ROOF ______________ Squares

KIND OF SHINGLES
- ☐ Cedar ☐ Cypress ☐ Pine
- ☐ Other ______________________

THICKNESS OF SHINGLES
- ☐ 4 butts–2 In. ☐ 5 butts–2 In.
- ☐ Other ______________________

EXPOSURE ______________ In.

SHINGLES PRESTAINED ☐ Yes ☐ No

FLASHINGS

Valley Flashings
- ☐ Metal — Kind ______________________
- ☐ Wood Shingles
- ☐ Other ______________________

Vent Flashings
- ☐ Metal — Kind ______________________
- ☐ Other ______________________

3.10: *(Continued)*

Drip Edge ☐ Metal Kind ____________________

☐ Wood Shingles ☐ Roll Roofing

Chimney Flashings ☐ Metal Kind ____________________

☐ Other ____________________

PREVIOUS MAINTENANCE (Describe Briefly with Dates)

Wood Shingles:

Flashings:

PREVIOUS REPAIRS (Describe Briefly with Dates)

Wood Shingles:

Flashings:

GENERAL REMARKS:

3.11: Asphalt Shingle Roofs—Historical Record

ASPHALT SHINGLE ROOFS—HISTORICAL RECORD

Form Completed By ______________________ Date ______________________

Building ______________________ Used for ______________________

☐ Permanent ☐ Temporary Year Roof Was Applied __________

TYPE OF ROOF DECK

☐ Sheathing Boards Thickness ______________

☐ Plywood Thickness ______

UNDERLAYER

☐ None ☐ Saturated Felt ☐ Paper

☐ Asphalt Shingles ☐ Wood Shingle

☐ Other ______________________

SLOPE OF ROOF ______________ In. per foot

AREA OF ROOF ______________ Squares

TYPE OF SHINGLES

☐ Strip Exposure ______________

☐ Individual Exposure ______________

Strip Shingles ☐ Square Butt ☐ Hexagonal

Square Butt Shingles ☐ Thick Butt ☐ Standard Weight

☐ Heavy Weight ☐ Class A

Hexagonal Shingles ☐ Standard Weight ☐ Heavy Weight

Individual Shingles ☐ Standard Weight ☐ Heavy Weight

Method of Laying ☐ American ☐ Dutch Lap ☐ Hexagonal

☐ Lock Down Exposure ______________

COLOR OF ROOFING GRANULES ______________________

FLASHINGS

Valley Flashings ☐ Roll Roofing ☐ Asphalt Shingles

☐ Metal Kind of Metal ______________

Drip Edge ☐ Roll Roofing ☐ Asphalt Shingles

☐ Metal Kind of Metal ______________

Vent Flashings ☐ Roll Roofing ☐ Metal

Kind of Metal ______________________

3.11: *(Continued)*

Chimney Flashings ☐ Roll Roofing ☐ Metal

Kind of Metal ______________________________

PREVIOUS MAINTENANCE (Describe Briefly with Dates)

Asphalt Shingles:

Flashings:

PREVIOUS REPAIRS (Describe Briefly with Dates)

Asphalt Shingles:

Flashings:

GENERAL REMARKS:

3.12: Painting—Historical Record

PAINTING—HISTORICAL RECORD

Prepared By ______________________ Date ______________

1. Facility ______________________
2. Year Building Was Previously Painted ______________________
3. Contract Number or Designation ______________________
4. Job Description ______________________

   ____
5. Paint Coats Applied ______________________
6. Area Painted ______________________
7. Gallons Used Of: ____________ Primer ____________ Paint
8. Total Man-hours Used ______________________
9. Condition of Existing Painted Surface Prior to Painting and Surface Preparation:

   ____
10. Brand Name of Primer Used and Type ______________________
11. Brand Name of Paint Used and Type ______________________
12. Weather Conditions During Painting ______________________
13. General Comments:

3.13: Building Paint Work Record

BUILDING PAINT WORK RECORD

Building Name	Bldg. No.	Area Sq. Yds.	Date Painted	P.O. No.	Paint Color and Type No.	Remarks

FORM 3.13

Key Use of Form:

To record/document all painting activities within the facility. Will use to determine when future painting activities need to be performed.

Who Prepares:

Plant engineering/maintenance personnel

Who Uses:

Plant engineering

How to Complete:

Provide the information requested in the appropriate spaces. Obtain from painting contractor and appropriate in-house personnel.

Alternative Forms:

Form can be revised to be company-specific by changing headings.

3.14: Glass Replacement Data

GLASS REPLACEMENT DATA

Date Reported	Date Replaced	No. of Lights	Location			Reason for Replacement	
			Floor	Elevation	Room No.	Broken Glass	Leakers

FORM 3.14

Key Use of Form:

To record (keep track of) when glass was replaced in the facility. Will be used to monitor glass replacement and location of same for purposes of resource control and potential functional problems within the facility.

Who Prepares:

Plant engineering/maintenance personnel.

Who Uses:

Plant engineering.

How to Complete:

Provide the information requested in the appropriate spaces. Obtain from maintenance and plant engineering staff.

Alternative Forms:

Form can be revised to be company-specific by changing the headings.

B. INSPECTION FORMS

The person making the inspection should first record all the general types of information requested such as Date of Inspection, Item Being Inspected, and so on. The next step is the inspection and recording of observations on this document, along with any suggested follow-up activities that should take place such as an inspection, a need for immediate repair, or a recommendation for replacement. More than one day may be needed to complete the inspection process.

After the form is completed, it should be transmitted to the designated maintenance personnel. In addition, copies should be kept on file for future reference. The information from these forms will be used to evaluate the existing maintenance program and make any needed corrections. Furthermore, the data becomes critical in that it provides helpful resource information prior to making any needed repairs such as when was the peeling of the paint first observed.

Forms 3.15 through 3.30 can be used to inspect interior and exterior building finishes. They are completed as other inspection forms.

3.15: Building Inspection Checklist (Interior and Exterior)

BUILDING INSPECTION CHECKLIST
(Interior and Exterior)

Maintenance Inspector's Name ____________________

Weather ____________________

EXTERIOR MAINTENANCE

Item to Be Inspected	Date	Comments
Roof Cracks		
Blisters (Vapor)		
Wrinkles		
Ponding		
Gravel Coming Off		
Wet Insulation		
Dry Felts		
Foreign Objects on Roof		
Other		
Flashing and Counterflashing Composition Flashing		
Cracks		
Open Joints or End Laps		
Pulled Away from Wall		
Other		
Metal Counterflashings Pulled Out of Reglets		
Loose Wedges		
Deteriorated Caulking		
Rusted Galvanized Iron		
General Deterioration		
Other		
Expansion Joints Cracks		
Mislocated or Loose Blocking		
Deteriorated Elastic Membrane		
Other		

3.15: *(Continued)*

Item to Be Inspected	Date	Comments
Gravel Stops and Gutters Peeling Paint		
Broken Solder Joints		
Separation from Roof Piles		
Other		
Skylights Secured to Supports		
Cracks		
Other		
Plumbing Vents Secured to Supports		
Cracks		
Other		
Antennas, Signs, and Other Similar Items Rust and Peeling Paint		
Loose Guys		
Roof Seal		
Other		
Parapets Open Mortar Joints		
Open Joints in Coping		
Other		
Chimneys Deteriorated Brick, Joints		
Flue Liner		
Flashing		
Other		
Walls Cracks		
Paint Condition		
Evidence of Moisture		
Efflorescence		
Areaways		

3.15: *(Continued)*

INTERIOR MAINTENANCE

Item to Be Inspected	Date	Comments
Walls *(Continued)* Caulking		
Other		
Doors Closers		
Exit Devices		
Locks and Misc. Hardware		
Deteriorated Surface		
Other		
Structural System Rot, Splitting, Rust		
Termites		
Deflection		
Cracks		
Other		
Stairs and Handrails Loose Handrails		
Worn Treads		
Paint or Varnish		
Other		
Interior Walls Cracks, Bulges		
Paint Conditions		
Dirt		
Other		
Ceilings Cracks, Bulges		
Paint Condition		
Dirt		
Other		

3.16: Building Rating Profile Summary Sheet

BUILDING RATING PROFILE SUMMARY SHEET

Facility ________________ Maintenance Inspector's Name ________________ Date of Inspection ________

Building Component	5 Excellent	4 Good	3 Average	2 Fair	1 Poor
Roof					
Building Exterior					
Parking Lots and Drives					
Landscaping					
Office Interiors					
Factory Interiors					
Heating/Ventilating/Air Cond.					
Plumbing					
Electrical Systems					
Lighting					
Fire Protection					
General Housekeeping					
Painting					

GENERAL REMARKS:

3.16: *(Continued)*

ANNUAL BUILDING INSPECTION

Detail Sheet

Facility ______________________________

Maintenance
Inspector's Name ______________________ Date of Inspection ____________

COMPONENT: Roof

APPRAISAL OF: Roofing Membrane, Flashings, Parapets, Drains, Gutters

CONDITION: Excellent - Good - Average - Fair - Poor (Circle One)

RECOMMENDATIONS:

COMPONENT: Building Exterior

APPRAISAL OF: Walls, Windows, Doors, Trim, Docks, Signs

CONDITION: Excellent - Good - Average - Fair - Poor (Circle One)

RECOMMENDATIONS:

COMPONENT: Parking Lots and Drives

APPRAISAL OF: Asphalt, Concrete, Curbs, Striping, Sidewalks

CONDITION: Excellent - Good - Average - Fair - Poor (Circle One)

RECOMMENDATIONS:

COMPONENT: Landscaping

APPRAISAL OF: Lawn, Trees, Shrubs, Retaining Walls, Fences

CONDITION: Excellent - Good - Average - Fair - Poor (Circle One)

RECOMMENDATIONS:

3.16: *(Continued)*

COMPONENT: Office Interiors

APPRAISAL OF: Walls, Ceilings, Floor Covering, Doors, Windows, Decor, Furniture

CONDITION: Excellent - Good - Average - Fair - Poor (Circle One)

RECOMMENDATIONS:

COMPONENT: Factory Interiors

APPRAISAL OF: Walls, Ceilings, Floors, Doors, Windows

CONDITION: Excellent - Good - Average - Fair - Poor (Circle One)

RECOMMENDATIONS:

COMPONENT: Heating–Ventilating–Air Conditioning

APPRAISAL OF: Units, Controls, Grilles, Diffusers, Boilers, Piping, Ducts, Conditions

CONDITION: Excellent - Good - Average - Fair - Poor (Circle One)

RECOMMENDATIONS:

COMPONENT: Plumbing

APPRAISAL OF: Piping, Fixtures, Water Heaters, Water Conditioning, Water Coolers

CONDITION: Excellent - Good - Average - Fair - Poor (Circle One)

RECOMMENDATIONS:

3.16: *(Continued)*

COMPONENT: Electrical Systems

APPRAISAL OF: Transformers, Switchgear, Panels, Bus-Ducts, Wiring

CONDITION: Excellent - Good - Average - Fair - Poor (Circle One)

RECOMMENDATIONS:

COMPONENT: Lighting

APPRAISAL OF: Fixtures, Lamps, Lenses, Light Level, Alignment, Cleanliness

CONDITION: Excellent - Good - Average - Fair - Poor (Circle One)

RECOMMENDATIONS:

COMPONENT: Fire Protection

APPRAISAL OF: Automatic Sprinkler System, Water Pressure, Extinguishers, Inspections

CONDITION: Excellent - Good - Average - Fair - Poor (Circle One)

RECOMMENDATIONS:

COMPONENT: General Housekeeping

APPRAISAL OF: Conditions, Standards, Methods, Program

CONDITION: Excellent - Good - Average - Fair - Poor (Circle One)

RECOMMENDATIONS:

3.16: *(Continued)*

COMPONENT: Painting

APPRAISAL OF: Exterior, Offices, Factory, Wall Coverings, Woodwork, Equipment

CONDITION: Excellent - Good - Average - Fair - Poor (Circle One)

RECOMMENDATIONS:

FORM 3.16

Key Use of Form:

To evaluate the condition of the various items listed. Also used to plan and schedule future maintenance and repair tasks.

Who Prepares:

Building maintenance inspector.

Who Uses:

Plant engineering.

How to Complete:

Provide the information requested in the appropriate spaces. The inspector uses his or her own judgment in assigning a specific condition.

Alternative Forms:

Form can be revised to be company-specific by changing the headings.

3.17: Survey of Building Exterior

SURVEY OF BUILDING EXTERIOR

Facility Being Inspected ____________________

Maintenance Inspector's Name ____________ Date of Inspection ____________

BUILDING

General Appearance	☐ Good	☐ Needs Improvement	
Walls	☐ Good	☐ Deteriorated	☐ Cracked
	☐ Evidence of Movement		☐ Repairs Needed
Mortar Joints	☐ Good	☐ Deteriorated	

Other ____________________

Canopies	☐ Good	☐ Repairs Required
Doors	☐ Good	☐ Repairs Required
Closers	☐ Good	☐ Repairs Required
Weatherstrips	☐ Good	☐ Repairs Required

Remarks:

SIDEWALKS

Surface	☐ Good	☐ Repairs Required	
Joints	☐ Good	☐ Open	
Movement	☐ Yes	☐ No	☐ Repairs Needed

Remarks:

PARKING LOT

Cleanliness	☐ Good	☐ Needs Improvement	
Striping	☐ Good	☐ Needs Improvement	
Curbs	☐ Good	☐ Movement	☐ Repairs Needed
Asphalt	☐ Good	☐ Repairs Required	

Remarks:

LANDSCAPING

General Appearance	☐ Good	☐ Needs Improvement
Irrigation System	☐ Operative	☐ Nonoperative

Remarks:

3.17: *(Continued)*

LIGHTING

Poles	☐ Good	☐ Rusty	☐ Movement
	☐ Repairs Needed	☐ Paint Needed	
Fixtures	☐ Good	☐ Rusty	
Lights	Quantity Operating ______		

Remarks:

SIGNS

General Appearance	☐ Good	☐ Needs Improvement
Illuminated Signs Operating	☐ Yes	☐ No

Remarks:

TRASH COMPACTOR–DUMPSTER

General Appearance	☐ Good	☐ Needs Improvement
Hopper	☐ Open	☐ Enclosed
Dead-Man Operation	☐ Yes	☐ No

Remarks:

MAJOR REPAIR–REPLACEMENT

Budgeted Current Year:

To Be Budgeted Next Year:

3.18: Survey of Building Interior

SURVEY OF BUILDING INTERIOR

Facility Being Inspected ______________________________

Maintenance Inspector's Name ______________ Date of Inspection ______________

COMPUTER ROOM

Cleanliness	☐ Good	☐ Needs Improvement	
Damped Mopped	☐ Daily	☐ Biweekly	☐ Weekly
A/C Filter Cleaned/Replaced	☐ Monthly	☐ Bimonthly	☐ Quarterly
Temperature ______________			
Alarm/Shutdown System	☐ Yes	☐ No	
Back-up A/C	☐ Yes	☐ No	
Equipment Logs Up To Date	☐ Yes	☐ No	

Remarks:

OFFICE AREA

Ceiling	☐ Good	☐ Needs Improvement
Floor	☐ Good	☐ Needs Improvement
Walls	☐ Good	☐ Needs Improvement
Carpet	☐ Good	☐ Needs Improvement
Equipment Maintained	☐ Yes	☐ No

Light Levels in Areas Indicated (Foot Candles) ______________ Mgr./Opr. Mgr.

__________ Personnel __________ Hallways __________ Controller

__________ Conference/Training Room

Remarks:

EATING AREA

Housekeeping	☐ Good	☐ Needs Improvement
Exhaust System	☐ Operating	☐ Not Operating
Exhaust System Cleaned/Treated		Date ______________
Filters	☐ Clean	☐ Needs Improvement
Equipment	☐ Clean	☐ Needs Improvement

Remarks:

3.18: *(Continued)*

RESTROOM

Cleanliness	☐ Good	☐ Needs Improvement
Fixtures	☐ Good	☐ Needs Repairs

Remarks:

STOCK AREAS

Overall Appearance	☐ Good	☐ Needs Improvement
Aisles Clear	☐ Yes	☐ No
Flues Open	☐ Yes	☐ No
Equipment Maintained	☐ Yes	☐ No
Doors/Hinges/Closers	☐ Good	☐ Needs Repairs

Light Levels (Foot Candles) __________ Marking Room __________ Stockroom

Remarks:

PAINTING AREA

Housekeeping	☐ Good	☐ Needs Improvement
Organization	☐ Good	☐ Needs Improvement
Paint Booth Area Maintained	☐ Yes	☐ No
Paint Booth Area Ventilated	☐ Yes	☐ No
Flammable Liquid Storage	☐ Acceptable	☐ Unacceptable

Light Level (Foot Candles) ________________

Remarks:

SAFETY-PROTECTION EQUIPMENT

Sprinklers Tested ☐ Yes Date ______________ ☐ No

Pressure __________

Fire Extinguisher Inspection Date ______________

No Smoking Observed in Appropriate Areas ☐ Yes ☐ No

Any Special Fire or Safety Hazards. If so, explain:

Remarks:

3.18: *(Continued)*

SALES AREA

Ceiling	☐ Good	☐ Needs Improvement	
Walls	☐ Good	☐ Needs Maintenance	
Hard Surface Floor	☐ Good	☐ Needs Improvement	
Build-Up	☐ Yes	☐ No	
Stripped	☐ Quarterly	☐ Semiannual	☐ Annual
Carpet	☐ Good	☐ Needs Improvement	
Cleaned	☐ Quarterly	☐ Semiannual	☐ Annual
Protective Coating Applied	☐ Yes	☐ No	

Date Last Cleaned ______________________

Fitting Rooms	☐ Good	☐ Needs Improvement

Light Levels (Foot Candles). List the locations and the light levels:

Remarks:

AUTOMOTIVE SHOP

Housekeeping	☐ Good	☐ Needs Improvement
Walls	☐ Good	☐ Needs Improvement

Light Level (Foot Candles) ________________

Eye Protection Available	☐ Yes	☐ No

Remarks:

BATTERY ROOM

Housekeeping	☐ Good	☐ Needs Improvement	
Walls	☐ Good	☐ Needs Improvement	
Charging Rack	☐ Good	☐ Corroded	
Buss Bars	☐ Good	☐ Corroded	
Leads	☐ Good	☐ Replacement Required	☐ Bolts Needed
Floor Protection	☐ Good	☐ Repairs Required	
Exhaust System	☐ Adequate	☐ Inadequate	
Protective Equipment Available		☐ Yes	☐ No

Remarks:

3.18: *(Continued)*

HOUSEKEEPING EQUIPMENT AND SUPPLIES

Equipment Maintained	☐ Yes	☐ No
Organized	☐ Yes	☐ No
Approved	☐ Yes	☐ No
Supplies Organized	☐ Yes	☐ No
Approved	☐ Yes	☐ No
Inventory Maintained	☐ Yes	☐ No
Equipment or Supplies from Unapproved Sources	☐ Yes	☐ No

Remarks:

MAJOR REPAIR–REPLACEMENT

Budgeted Current Year:

To Be Budgeted Next Year:

GENERAL NOTES:

3.19: Roof Inspection Checklist

ROOF INSPECTION CHECKLIST

Maintenance Inspector's Name ______________________________

Description of Items Being Inspected ______________________________

____

Instructions: Perform the specified inspection tasks. Insert the date when the activity is complete. Make any comments which are pertinent to future maintenance needs.

Item to Be Inspected	Date	Comments
Inspect wood shingles for weathering; warped, broken, split; curling; missing; flashing failures.		
Inspect tile for weathering; broken, cracked, loose, missing; flashing failures; deterioration of expansion-joint material or tile raising.		
Inspect slate for weathering; broken, cracked, loose, missing; flashing failures.		
Inspect metal for holes, looseness, punctures, broken seams, adequate side and end laps, adequate expansion joints, rust or corrosion and damage.		
Inspect cement-composition roofing for wear from weathering, broken, cracked, loose, missing, sufficient side or end lap.		
Inspect asphalt-roll roofing for weathering, cracking, alligatoring, buckling, blistering, sufficient laps, tearing, other damage to coatings.		
Inspect asphalt-shingle roofing for lifting, weathering, cracking, curling, buckling, blistering, loss of granules.		
Inspect built-up roofing for cracking, alligatoring, low spots and water ponding; failure or lack of gravel stops; cracks in membrane; exposed bituminous coatings; exposed, disintegrated, curled or buckled felts.		
Inspect fastenings for proper, loose, missing, broken, defective, weathered materials.		

3.19: *(Continued)*

Item to Be Inspected	Date	Comments
Inspect metal base flashings for rust vertical joints; flanges, adequate nailing; proper fastening; proper sealing; cant strip.		
Inspect other base flashings for sagging; separation; adequate coverage or embedment; vertical joints; proper fastening; buckling; cracking; surface coat, cant strip.		
Inspect cap flashings for open joints, buckling, cracking; surface coat; proper fastenings, rust, and corrosion, as applicable.		
Inspect chimney, wall, ridge, vent, valley, and edge flashings for open joints, loose, proper fastenings, other damage.		
Inspect concrete slabs for cracks, spalling, expansion joints, low spots, drainage.		
Inspect parapet walls and copings for cracks, spalling, joints, other damage.		

Provide any suggestions or recommendations for follow-up activities or any other comments related to the inspection and maintenance activities.

3.20: Buildings Inspection Checklist (Except Roofs and Trusses)

BUILDINGS INSPECTION CHECKLIST
(Except Roofs and Trusses)

Maintenance Inspector's Name ____________________

Description of Items Being Inspected ____________________

____

Instructions: Perform the specified inspection tasks. Insert the date when the activity is complete. Make any comments which are pertinent to future maintenance needs.

Item to Be Inspected	Date	Comments
Check concrete for spalling, breaks, exposed reinforcing, settlement or buckling, and damage.		
Check masonry for eroded or sandy mortar joints, leaks, soft or spalling bricks, settlement or buckling, and damage.		
Check timber for warping, checking, splitting, bowing, sagging, deflection, rotting, loose, damaged, or missing bolts, split rings and other connections.		
Check building for termite or insect infestation.		
Check for proper drainage.		
Check walls for stains, holes, sagging, buckling, support failure, open joints, rust, corrosion, rot, and damage, as applicable.		
Check wall coverings for curling, looseness, punctures, fading, wrinkling, stains, and damage.		
Check screens for binding, poor fit of frames, holes in fabric, and damage.		
Check screen frames for rust, corrosion, rot, and damage, as required.		
Check doors and windows for looseness, missing sealant, poor fit, rust, corrosion, rot, damage, and broken or missing glass.		
Check floors and stairs for sagging, wear, warping, rotting, stains, and damage.		

3.20: *(Continued)*

Item to Be Inspected	Date	Comments
Check floor coverings for wear, cracking, chipping, tears, cuts, raveling, stains, fading, failure to bond, and other damage, as applicable.		
Check paint for alligatoring, checking, blistering, scaling, peeling, wrinkling, fading, chalking, mildew, bleeding, stains, and failure of bond.		
Check gutters, and downspouts for misalignment, rust, corrosion, clogging, leaks, missing guards and fastenings, and damage.		

Provide any suggestions or recommendations for follow-up activities or any other comments related to the inspection and maintenance activities.

3.21: Survey of Roof Condition

SURVEY OF ROOF CONDITION

Facility ______________________ Location ______________________

Maintenance Inspector's Name ______________ Date of Inspection ______________

OVERALL CONDITION

☐ Good ☐ Fair ☐ Poor

ROOF SYSTEM

Ponding ☐ Yes ☐ No

If yes, does it evaporate within 48 hours?

☐ Yes ☐ No

Note areas that do not evaporate within 48 hours on the roof detail grid.

Damage ☐ Traffic ☐ Severe Weather ☐ Storage

☐ Other ______________________

Coverage of Mineral or Aggregate ☐ Good ☐ Poor

Adhesion of Mineral or Aggregate ☐ Good ☐ Poor

Surface Bituminous Coating ☐ Good ☐ Cracked

☐ Other ______________________

Membrane Exposed ________ % Deteriorated ________ %

☐ Cracks ☐ Fishmouths ☐ Blisters

☐ Ridging ☐ Other ______________

Base Flashing ☐ Good ☐ Deterioration ☐ Joints Open

☐ Slipping from Parapet ☐ Wrinkled

☐ Cracked ☐ Other ______________

Cap-Counter Flashing ☐ Good ☐ Open Joints

☐ Loose ☐ Corroded

☐ Other ______________________

Gravel Stops ☐ Good ☐ Loose ☐ Corroded

☐ Joints Open ☐ Other ______________

Flashed Pipes—Vents ☐ Good ☐ Repairs Needed

Hooded Penetration ☐ Good ☐ Repairs Needed

Pitch Pockets—Pans ☐ Good ☐ Cracked

☐ Low ☐ Other ______________

DRAINAGE SYSTEM

Scuppers ☐ Good ☐ Deterioration ☐ Joints Open

☐ Other ______________________

3.21: *(Continued)*

Gutters	☐ Good	☐ Deteriorated	☐ Joints Open
	☐ Other ________		
Downspouts	☐ Good	☐ Deteriorated	
	☐ Other ________		
Drains	☐ Open	☐ Clogged	
PARAPET WALLS—COPING			
Mortar	☐ Good	☐ Deteriorated	
Plaster	☐ Good	☐ Deteriorated	☐ Cracks
Metal Coping	☐ Good	☐ Paint Peeling	☐ Corroded
	☐ Joints Open	☐ Other ________	
Masonry Coping	☐ Good	☐ Deteriorated	☐ Cracked
	☐ Joints Open	☐ Other ________	

REMARKS:

FORM 3.21

Key Use of Form:

To evaluate condition of roof. Also used to plan and schedule maintenance and/or repair activities on the roof.

Who Prepares:

Maintenance inspector

Who Uses:

Plant engineering

How to Complete:

Provide the information requested in the appropriate spaces.

Alternative Forms:

Form can be revised to be company-specific by changing the items.

3.21: *(Continued)*

REQUIRED REPAIRS			
Area	**Routine**	**Immediate**	**Description**

MAJOR REPAIR—REPLACEMENT

Budgeted Current Year ______________________________

To Be Budgeted Next Year ______________________________

3.21: *(Continued)*

ROOF DETAIL GRID

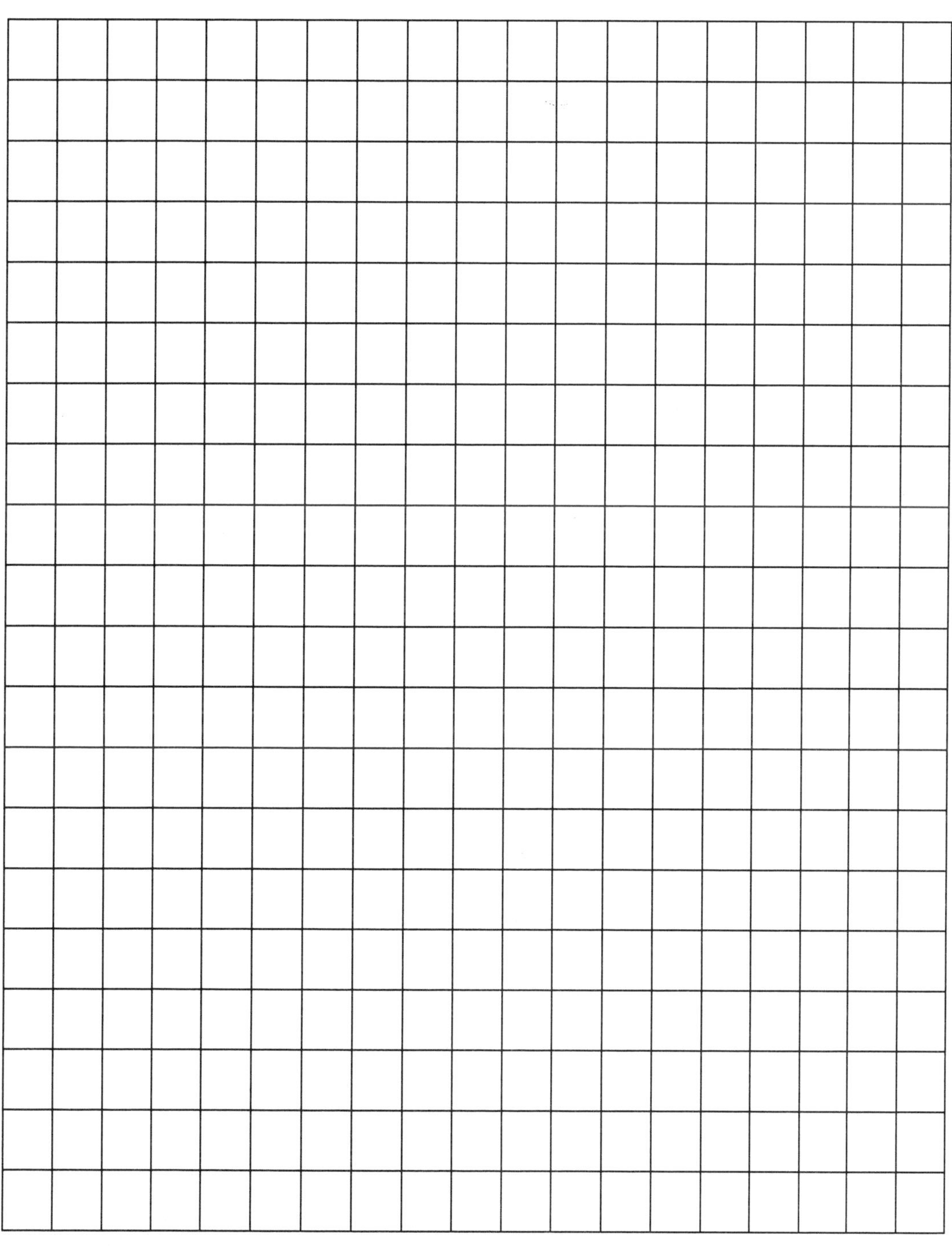

3.22: Condition of Roof Survey Report

CONDITION OF ROOF SURVEY REPORT

Facility ____________________ Location ____________________ Roof Area __________

Building Use ________________________ Roof Designation ________________________

Constructed ____________________ Age ____________________

Maintenance Inspector's Name ______________________ Date of Inspection __________

Specification:

Present Condition:

Recommendations:

General Remarks:

FORM 3.22

Key Use of Form:

To evaluate condition of roof. Used also to plan and schedule maintenance and/or repair activities on the roof.

Who Prepares:

Maintenance inspector

Who Uses:

Plant engineering

How to Complete:

Record the results of the inspection in the spaces provided. Obtain information from drawings, specifications, and field observation.

Alternative Forms:

Forms can be made company-specific by making appropriate revisions to it.

3.23: Wood Shingle Roof Inspection Checklist

WOOD SHINGLE ROOF INSPECTION CHECKLIST

Building ______________________ Location ______________________

Maintenance Inspector's Name ______________ Date of Inspection ______________

GENERAL APPEARANCE

	☐ Good	☐ Fair	☐ Poor
Watertightness	☐ No Leaks	☐ Leaks with Long Rain	
	☐ Leaks with Every Rain		

REPORTED CAUSE OF LEAKS

☐ Weathering	☐ Cracked Shingles	☐ Curled Shingles
☐ Failure of Nails	☐ Flashing Failure	☐ Other ________

CONDITION OF SHINGLES

☐ Unchanged	☐ Cracked ________ %	☐ Curled ________ %
☐ Loose ________ %		

GENERAL CONDITION OF WOOD SHINGLE ROOF:

TREATMENT RECOMMENDED:

FLASHINGS		
Chimney Flashings	☐ Satisfactory	☐ Defective
Wall Flashings	☐ Satisfactory	☐ Defective
Ridge Flashings	☐ Satisfactory	☐ Defective
Vent Flashings	☐ Satisfactory	☐ Defective
Valley Flashings	☐ Satisfactory	☐ Defective
Edge Flashings	☐ Satisfactory	☐ Defective
DRAINAGE SYSTEM		
Gutters	☐ Satisfactory	☐ Defective
Downspouts	☐ Satisfactory	☐ Defective

Treatment Recommended:

GENERAL NOTES:

3.24: Built-Up Roof Inspection Checklist

BUILT-UP ROOF INSPECTION CHECKLIST

Building ______________________ Location ______________________

Maintenance Inspector's Name ______________ Date of Inspection ______________

ROOFING MEMBRANE

General Appearance ☐ Good ☐ Fair ☐ Poor

Watertightness ☐ No Leaks ☐ Leaks with Long Rain ☐ Leaks with Every Rain

REPORTED CAUSE OF LEAKS

☐ Weathering of Material ☐ Faulty Material
☐ Faulty Design ☐ Wind Damage ☐ Hail Damage
☐ Faulty Construction ☐ Roof Traffic
☐ Flashing Failure ☐ Low Spots
☐ Gravel Stop Failure
☐ Other ______________________

ADHESION OF MINERAL SURFACING TO BITUMEN

☐ Good ☐ Fair ☐ Poor

BARE AREAS (Give approximate percentage of total roof area below.)

Bituminous Coating Exposed __________ %

Condition of Coating ☐ Smooth ☐ Alligatored ☐ Cracked

Condition of Felts ☐ Exposed ☐ Disintegrated ☐ Edges Curled

Blisters __________ (Give size, range and approximate number per square.)

Cracked to allow water to enter ☐ Yes ☐ No

Buckles ☐ Yes ☐ No

Cracked to allow water to enter ☐ Yes ☐ No

Cracks in membrane through to roof deck ☐ Yes ☐ No

Fishmouths ☐ Yes ☐ No

GENERAL CONDITION OF THE ROOF MEMBRANE:

TREATMENT RECOMMENDED:

3.24: *(Continued)*

FLASHINGS

BASE FLASHINGS

Metal: ☐ Deteriorated ☐ Vertical Joints Open

Flanges of Base Metal Flashing Loose ☐ Yes ☐ No

Caused By ☐ Inadequate Nailing

☐ Not Properly Sealed with Felt Strips

Plastic:

☐ Sagged or Separated from Parapet Wall

☐ Buckled ☐ Cracked

Failure of Base Flashing ☐ Weathering ☐ Mechanical

Surface Coating Disintegrated ☐ Yes ☐ No

Vertical Laps Not Cemented Properly ☐ Yes ☐ No

CAP FLASHINGS

Metal:

Firmly Embedded into Vertical Wall ☐ Yes ☐ No

Flashing ☐ Deteriorated ☐ Vertical Joints Open

Not Covering Base Flashing Adequately ☐ Yes ☐ No

Plastic:

Surface Coating Disintegrated ☐ Yes ☐ No

Flashing Felt Disintegrated ☐ Yes ☐ No

FLASHING BLOCK

Groove Pointed Sufficiently ☐ Yes ☐ No

Recommended Treatment:

PARAPET WALLS

Mortar Joints Deteriorated ☐ Yes ☐ No

Settlement Cracks in Walls ☐ Yes ☐ No

Joints in Tile Coping Open ☐ Yes ☐ No

Concrete Coping Cracked ☐ Yes ☐ No

Other Defects ☐ Describe ______________________________

____

Recommended Treatment:

GENERAL NOTES:

3.25: Asphalt Shingle Roof Inspection Checklist

ASPHALT SHINGLE ROOF INSPECTION CHECKLIST

Building ______________________ Location ______________________

Maintenance Inspector's Name ______________ Date of Inspection ______________

ASPHALT SHINGLES

General Appearance	☐ Good	☐ Fair	☐ Poor
Watertightness	☐ No Leaks	☐ Leaks with Long Rain	
	☐ Leaks with Every Rain		
Reported Cause of Leaks	☐ Wind	☐ Weathering of Shingles	
	☐ Faulty Material	☐ Faulty Design	
	☐ Faulty Application		
Other Problems	☐ Hail Damage	☐ Traffic on Roof	
	☐ Other Mechanical Damage ______________		
	☐ Failure of Flashings		

CONDITION OF SHINGLES

	☐ Unchanged	☐ Buckled	☐ Blistered
	☐ Curled	☐ Tabs Missing	
Loss of Granules	☐ Slight	☐ Medium	☐ Severe
Other Problems	☐ Asphalt Coating Damaged		
	☐ Coating Alligatored or Cracked		
	☐ Other ______________		

GENERAL CONDITION OF ROOF:

TREATMENT RECOMMENDED:

FLASHINGS

Chimney Flashings	☐ Satisfactory	☐ Defective
Wall Flashings	☐ Satisfactory	☐ Defective
Ridge Flashings	☐ Satisfactory	☐ Defective
Vent Flashings	☐ Satisfactory	☐ Defective
Valley Flashings	☐ Satisfactory	☐ Defective
Edge Flashings	☐ Satisfactory	☐ Defective

DRAINAGE SYSTEM

Gutters	☐ Satisfactory	☐ Defective
Downspouts	☐ Satisfactory	☐ Defective

GENERAL REMARKS:

3.26: Cement Composition Roof Inspection Checklist

CEMENT COMPOSITION ROOF INSPECTION CHECKLIST

Building ______________________ Location ______________________

Maintenance Inspector's Name ______________ Date of Inspection ______________

CEMENT COMPOSITION ROOFING

General Appearance ☐ Good ☐ Fair ☐ Poor

Watertightness ☐ No Leaks ☐ Leaks with Long Rain ☐ Leaks with Every Rain

Reported Cause of Leaks ☐ Weathering ☐ Faulty Construction ☐ Faulty Design ☐ Faulty Material ☐ Wind ☐ Hail ☐ Flashings ☐ Underlayment

☐ Other ______________________

CONDITION OF ROOFING

☐ Unchanged ☐ Loose Shingles

☐ Broken Shingles ______________ %

☐ Broken Corrugated Sheets __________ %

Failure of Fasteners ☐ Yes ________ % ☐ No

GENERAL CONDITION OF ROOF:

TREATMENT RECOMMENDED:

FLASHINGS

Chimney Flashings	☐ Satisfactory	☐ Defective
Wall Flashings	☐ Satisfactory	☐ Defective
Ridge Flashings	☐ Satisfactory	☐ Defective
Vent Flashings	☐ Satisfactory	☐ Defective
Valley Flashings	☐ Satisfactory	☐ Defective
Edge Flashings	☐ Satisfactory	☐ Defective

DRAINAGE SYSTEM

Gutters	☐ Satisfactory	☐ Defective
Downspouts	☐ Satisfactory	☐ Defective

GENERAL REMARKS:

3.27: Metal Roof Inspection Checklist

METAL ROOF INSPECTION CHECKLIST

Building ______________________ Location ______________________

Maintenance Inspector's Name ______________ Date of Inspection ______________

TYPE OF METAL: ____________________

General Appearance	☐ Good	☐ Fair	☐ Poor
Watertightness	☐ No Leaks	☐ Leaks with Long Rain	
	☐ Leaks with Every Rain		
Reported Causes of Leaks	☐ Corrosion	☐ Faulty Design	☐ Broken Seams
	☐ Faulty Construction		☐ Faulty Seams
	☐ Insufficient Lap	☐ Defective Fasteners	
	☐ Flashings	☐ Other ______________	

GENERAL CONDITION

Rust or Corrosion	☐ None	☐ Slight	☐ Severe
Condition of Protective Coating	☐ Good	☐ Fair	☐ Poor
Seams Broken	☐ Yes	☐ No	
	☐ Location ______________		
Other Breaks	☐ Yes	☐ No	
	☐ Location ______________		
Holes	☐ Yes	☐ No	
	☐ Location ______________		
Expansion Joints	☐ Sufficient No.		☐ Too Few
	☐ Indicate Where Needed ______________		

GENERAL CONDITION OF ROOF:

TREATMENT RECOMMENDED:

3.27: *(Continued)*

FLASHINGS		
Chimney Flashings	☐ Satisfactory	☐ Defective
Wall Flashings	☐ Satisfactory	☐ Defective
Ridge Flashings	☐ Satisfactory	☐ Defective
Vent Flashings	☐ Satisfactory	☐ Defective
Valley Flashings	☐ Satisfactory	☐ Defective
Edge Flashings	☐ Satisfactory	☐ Defective
DRAINAGE SYSTEM		
Gutters	☐ Satisfactory	☐ Defective
Downspouts	☐ Satisfactory	☐ Defective

GENERAL REMARKS:

3.28: Slate Roof Inspection Checklist

SLATE ROOF INSPECTION CHECKLIST

Building ______________________________ Location ______________________________

Maintenance Inspector's Name ________________ Date of Inspection ________________

SLATE ROOFING

General Appearance	☐ Good	☐ Fair	☐ Poor
Watertightness	☐ No Leaks	☐ Leaks with Long Rain	
	☐ Leaks with Every Rain		
Reported Cause of Leaks	☐ Weathering	☐ Faulty Material	
	☐ Faulty Construction		☐ Faulty Design
	☐ Wind	☐ Hail	☐ Flashings
	☐ Underlayer	☐ Other ____________	

CONDITION OF SLATE

	☐ Unchanged	☐ Disintegrated	
Amount of Disintegration	☐ Slight	☐ Severe	☐ Broken
	☐ Other ____________		
Failure of Fasteners	☐ No	☐ Yes ______ %	

GENERAL CONDITION OF ROOF:

TREATMENT RECOMMENDED:

FLASHINGS

Chimney Flashings	☐ Satisfactory	☐ Defective
Wall Flashings	☐ Satisfactory	☐ Defective
Ridge Flashings	☐ Satisfactory	☐ Defective
Vent Flashings	☐ Satisfactory	☐ Defective
Valley Flashings	☐ Satisfactory	☐ Defective
Edge Flashings	☐ Satisfactory	☐ Defective

DRAINAGE SYSTEM

Gutters	☐ Satisfactory	☐ Defective
Downspouts	☐ Satisfactory	☐ Defective

GENERAL REMARKS:

3.29: Tile Roof Inspection Checklist

TILE ROOF INSPECTION CHECKLIST

Building ______________________ Location ______________________

Maintenance Inspector's Name ______________ Date of Inspection ______________

TILE ROOFING

General Appearance	☐ Good	☐ Fair	☐ Poor
Watertightness	☐ No Leaks	☐ Leaks with Long Rain	
	☐ Leaks with Every Rain		
Reported Cause of Leaks	☐ Weathering	☐ Faulty Design	☐ Wind
	☐ Faulty Construction		☐ Hail
	☐ Faulty Material	☐ Flashings	
	☐ Underlayer	☐ Other ______________	

CONDITION OF TILE

	☐ Unchanged	☐ Broken Tiles __________ %
Cause of Breakage	☐ Nailed Too Tightly	
	☐ Other ______________________	
Failure of Fasteners	☐ No	☐ Yes ________ %
Other Failures	☐ Describe ______________________	

GENERAL CONDITION OF ROOF:

TREATMENT RECOMMENDED:

FLASHINGS

Chimney Flashings	☐ Satisfactory	☐ Defective
Wall Flashings	☐ Satisfactory	☐ Defective
Ridge Flashings	☐ Satisfactory	☐ Defective
Vent Flashings	☐ Satisfactory	☐ Defective
Valley Flashings	☐ Satisfactory	☐ Defective
Edge Flashings	☐ Satisfactory	☐ Defective

DRAINAGE SYSTEM

Gutters	☐ Satisfactory	☐ Defective
Downspouts	☐ Satisfactory	☐ Defective

GENERAL REMARKS:

3.30: Asphalt Roll Roofing Roof Inspection Checklist

ASPHALT ROLL ROOFING ROOF INSPECTION CHECKLIST

Building ______________________ Location ______________________

Maintenance Inspector's Name ______________ Date of Inspection ______________

ASPHALT ROLL ROOFING

General Appearance	☐ Good	☐ Fair	☐ Poor
Watertightness	☐ No Leaks	☐ Leaks with Long Rain	
	☐ Leaks with Every Rain		
Reported Cause of Leaks	☐ Weathering	☐ Faulty Material	☐ Wind
	☐ Faulty Application	☐ Hail Damage	☐ Traffic
	☐ Other ______________		

CONDITION OF ROOFING

	☐ Unchanged	☐ Buckled	☐ Blistered
Loss of Granules	☐ Slight	☐ Medium	☐ Severe
Other Problems	☐ Asphalt Coating Damaged		
	☐ Coating Alligatored or Cracked		
	☐ Other ______________		

GENERAL CONDITION OF ROOF:

TREATMENT RECOMMENDED:

FLASHINGS

Chimney Flashings	☐ Satisfactory	☐ Defective
Wall Flashings	☐ Satisfactory	☐ Defective
Ridge Flashings	☐ Satisfactory	☐ Defective
Vent Flashings	☐ Satisfactory	☐ Defective
Valley Flashings	☐ Satisfactory	☐ Defective
Edge Flashings	☐ Satisfactory	☐ Defective

DRAINAGE SYSTEM

Gutters	☐ Satisfactory	☐ Defective
Downspouts	☐ Satisfactory	☐ Defective

GENERAL REMARKS:

Section 4

LANDSCAPING AND GROUNDS FACILITIES

This section includes forms that can be used to inventory items relating to landscaping and grounds facilities (4.1 through 4.3). In addition, formats are included for the inspection and maintenance of landscaping and grounds (4.4 through 4.22).

For formats that can be used for planning, estimating, scheduling, ordering and evaluating inspection, maintenance, and repair activities on landscaping items and grounds facilities, refer to Section 8, *Maintenance Management*.

A. INVENTORY FORMS

The first series of forms (4.1 through 4.3) are to be completed by the appropriate maintenance management personnel to inventory vegetation (i.e., grasses, trees) and grounds facilities such as fences, retaining walls, check dams, and others. These forms serve as the foundation to an effective maintenance control system.

There should be one form completed for each type of vegetation and grounds facilities. As inspections, repairs, and other maintenance activities are performed on each item, the date accomplished, along with other pertinent facts such as maintenance activities performed, should be recorded on the inventory form. The person who will complete the form will depend on the type and size of the company.

This form is open-ended in that information will continue to be added to it. It serves as the historical basis for all maintenance aspects of the vegetation and grounds facilities. If problems arise, you can refer to the forms to first determine what kinds of maintenance activities have occurred prior to taking any follow-up action. This will result in a more efficient plan of action to solve the problem.

4.1: Grasses, Trees, and Shrubbery Inventory

GRASSES, TREES, AND SHRUBBERY INVENTORY

Facility Name ______________________ Location ______________________

Reference Drawing Numbers* ______________________

Vegetation Identification and Location ______________________

____

Instructions: Briefly describe the species of grass, trees, and shrubs, and the method of installation:

Frequency of Inspection ______________________

Responsibility of Inspection ______________________

INSPECTION DATA

Inspection Performed By	Date	Comments

*Two sets of Landscape Drawings and Specifications should be on file.

4.1: *(Continued)*

REPAIR (MAINTENANCE) DATA

Instructions: Briefly describe the regular maintenance activities that must be performed on the noted vegetation:

Inspection Performed By	Date	Work Performed

OTHER COMMENTS:

4.2: Exterior Ground Facilities Inventory

EXTERIOR GROUND FACILITIES INVENTORY

Facility Name ______________________ Location ______________

Reference Drawing Numbers* ______________________

Facility Designation and Location ______________________

Frequency of Inspection ______________________

Responsibility of Inspection ______________________

INSPECTION DATA

Inspected By	Date	Comments

*Two sets of Landscape Drawings and Specifications should be on file.

4.2: *(Continued)*

REPAIR (MAINTENANCE) DATA

Performed By	Date	Work Performed

OTHER COMMENTS:

4.3: Asphalt Surface Record

ASPHALT SURFACE RECORD

Completed by ______________________

Area Location	Area Size	Total Sq. Ft.	Composition of Surface	Schedule Work Priority				Activity				Remarks
				1	2	3	4	Crack Fill	Seal Coat	Patch	Surface	

FORM 4.3

Key Use of Form:

To document maintenance and repair activities performed on asphalt surfaces. Also used to plan and schedule appropriate repair of asphalt.

Who Prepares:

Plant engineering/maintenance personnel.

Who Uses:

Plant engineering and maintenance supervision.

How to Complete:

Using information obtained from the field (dimensions) and drawings provide the requested information in the appropriate spaces. Also, have plant engineering place a priority on the work to be done.

Alternative Forms:

A company-specific form similar to this one can be prepared by making the appropriate revisions to the existing form.

B. INSPECTION FORMS

The next series of forms (4.4–4.15) are to be used for inspecting the more common types of grounds facilities.

The person making the inspection should first record all the general types of information requested such as Date of Inspection, Item Being Inspected, and so on. The next step is the inspection and recording of observations on this document, along with any suggested follow-up activities that should take place such as an inspection, a need for immediate repair, or a recommendation for replacement. More than one day may be needed to complete the inspection process.

After the form is completed, it should be transmitted to the designated maintenance personnel. In addition, copies should be kept on file for future reference.

4.4: Grounds Inspection Checklist

GROUNDS INSPECTION CHECKLIST

Maintenance Inspector's Name ______________________________

Description and Location of Items Being Inspected ______________________________

____

Instructions: Perform the specified inspection tasks. Insert the date when the activity is complete. Make any comments which are pertinent to future maintenance needs.

Item to Be Inspected	Date	Comments
Inspect lawn and other turf areas including borders for traffic damage; color; density; sparse and bare spots; weeds; undesirable grasses; diseases; insect damage; erosion; silt deposits; waterborne debris; excessive height.		
Inspect trees and shrubs in landscaped areas for vigor, need of trimming, interference with utilities or buildings, injury from mowers, structural weaknesses, and storm, disease, or insect damage.		
Check border strips and areas seeded to rough grasses for poisonous or noxious weeds; seedling trees that may hinder future mowing; erosion and siltation; lack of vigor; inadequacy of coverage; evidence of burning.		
Check woodlands for erosion; dead, diseased, or damaged trees; firelanes for ease of access; vegetation growth that may carry ground fires; hollow trees.		
Check earth dams and dikes for damage from erosion, burrowing animals; seepage; lack of vegetation or vigor of growth; drop inlet pipes for stoppage; logs, debris; outlet ends for erosion, seepage, piping damage or failure.		
Check emergency spillways of drop inlet dams for blockage, erosion damage.		
Check permanent check-dams in water course for overflow at notch section, bypassing at ends, erosion on downstream side, damaged and deteriorated walls and apron.		

4.4: *(Continued)*

Item to Be Inspected	Date	Comments
Check hillside and terrace diversion embankment, channels, and culverts for silt, debris, rank vegetation, low and weak sections, overflow, erosion, gullying, burrowing animals.		
Check valley drainage channels including culverts and lateral drains and tile at entrance points for overflow, stoppage, silt, debris, rank vegetation, erosion, caving, sloughing, scour.		
Check vegetated waterways for adequate vegetation fullness and cover in relation to ground surface area that should be shielded; erosion of waterway and along sides; debris; overflow.		
Check fill slopes on barricades, highways, railways, airfield runways, igloos, and other soil-covered buildings for erosion, burning, steepness; lack of vigor and insufficient vegetation coverage for protection against beating rain and direct sunshine; inadequate fill depth at top of slope wherever buildings and weather conditions necessitate variations on different slopes; inadequate surface runoff piping; insufficient thickness of inorganic mulch (gravel, slag).		
Check cut slopes and diversion channels for erosion, scour, burning, weaknesses from past or possible overflow, lack of vigor or growth and insufficient vegetation coverage; inadequate surface runoff piping.		
Check gulleys including all surface water entrances and upstream ends or head where mainstream enters for current rate of erosion; resulting pollution and sedimentation of downstream lakes, channels; damaged lands; impairment of bridges and other structures; need of erosion control such as temporary brush and wire dams and plantings.		

4.4: *(Continued)*

Item to Be Inspected	Date	Comments
Check sprinkler system nozzles, sprays, hose, pipe, and valves for rust, corrosion, clogging, inadequate width or pressure, leakage, defective operation, evidence of water usage waste.		
Check flood irrigation systems including delivery channels, gates, flow control and water turnout works, and border dikes for defective operation, erosion, silting, scour, water loss, improper application, failure to supply to all parts of tract.		

Provide any suggestions or recommendations for follow-up activities or any other comments related to the inspection and maintenance activities.

4.5: Pavement Inspection Checklist

PAVEMENT INSPECTION CHECKLIST

Maintenance Inspector's Name ______________________________

Description and Location of Items Being Inspected ______________________________

____

Instructions: Perform the specified inspection tasks. Insert the date when the activity is complete. Make any comments which are pertinent to future maintenance needs.

Item to Be Inspected	Date	Comments
Check curbs, gutters for cracks, breaks, alignment, damaged tops, and adequacy of expansion joints.		
Check expansion joints for sufficient joint filler, proper bonding of filler, and for foreign material in joints.		
Check rigid pavements for spalling, cracks, depressions, scaling, buckling, and damage.		
Check flexible pavements for grooving, shoving, raveling, bleeding, burned areas, depressions, and damage.		
Check brick and stone for depressions, loose or missing parts, and grout or bedding failure.		
Check gravel, cinder, shell, and stabilized soil for breaks, potholes, deterioration, and damage.		
Check markings and signs for legibility.		

Provide any suggestions or recommendations for follow-up activities or any other comments related to the inspection and maintenance activities.

4.6: Cathodic Protection System Inspection Checklist

CATHODIC PROTECTION SYSTEM INSPECTION CHECKLIST

Maintenance Inspector's Name ____________________

Description and Location of Items Being Inspected ____________________

____

Instructions: Perform the specified inspection and maintenance tasks. Insert the date when the activity is complete. Make any comments which are pertinent to future maintenance needs.

Item to Be Inspected	Date	Comments
Check terminals and permanently installed test lead jumpers, accessible on underground systems, for rust, corrosion, broken or frayed wiring, loose connections and other deficiencies. Clean, tighten, repair, or replace as required.		
Check anode suspensions (elevated water tanks and systems for waterfront structures) for rust, corrosion, bent or broken suspension members or braces, frayed or broken suspension lines or cables, loose bolts, loose cable connections, frayed or broken wiring. Clean, repair, tighten, or replace as required.		
Check bushing (supporting anode) for rust and corrosion, broken or frayed wires, and loose connections. Tighten, repair, or replace as required.		
Record voltmeter reading.		
Record ammeter reading.		
Inspect wiring and electrical controls for loose connections; charred, broken, or wet insulation; evidence of short-circuiting and other deficiencies. Tighten, repair, or replace as required.		

Provide any suggestions or recommendations for follow-up activities or any other comments related to the inspection and maintenance activities.

4.7: Retaining Wall Inspection Checklist

RETAINING WALL INSPECTION CHECKLIST

Maintenance Inspector's Name ______________________________

Description and Location of Items Being Inspected ______________________________

____

Instructions: Perform the specified inspection tasks. Insert the date when the activity is complete. Make any comments which are pertinent to future maintenance needs.

Item to Be Inspected	Date	Comments
Inspect concrete foundation and concrete or masonry walls for spalling, cracks, and other evidence of deterioration or damage.		
Inspect timber walls and cribbing for rot, insect infestation, and other evidence of deterioration.		
Inspect sheet piling and bulkheads for rust, corrosion, bulging, and alignment.		
Check expansion joints condition.		
Check condition of attachments and fastenings.		
Inspect embankment slopes and area behind wall for erosion, settlement, and other damage.		

Provide any suggestions or recommendations for follow-up activities or any other comments related to the inspection and maintenance activities.

4.8: Seawall and Breakwater Inspection Checklist

SEAWALL AND BREAKWATER INSPECTION CHECKLIST

Maintenance Inspector's Name ____________________

Description and Location of Items Being Inspected ____________________

____

Instructions: Perform the specified inspection tasks. Insert the date when the activity is complete. Make any comments which are pertinent to future maintenance needs.

Item to Be Inspected	Date	Comments
Check for horizontal and vertical alignment.		
Check area around seawall for erosion.		
Check adequacy of riprap and rubble mound.		
Check cellular type breakwater for adequacy of sand or rock fill.		
Check curbing, handrails, and catwalks for loose, missing, or broken sections, obstructions, and other hazardous conditions.		

Provide any suggestions or recommendations for follow-up activities or any other comments related to the inspection and maintenance activities.

4.9: Tunnel and Underground Structure Inspection Checklist

TUNNEL AND UNDERGROUND STRUCTURE INSPECTION CHECKLIST

Maintenance Inspector's Name ______________________________

Description and Location of Items Being Inspected ______________________________

____

Instructions: Perform the specified inspection tasks. Insert the date when the activity is complete. Make any comments which are pertinent to future maintenance needs.

Item to Be Inspected	Date	Comments
Comply with all current safety precautions.		
Inspect structures for proper drainage, cracks, breaks, leaks in face and between face and tunnel lining, eroded slopes, or undermining.		
Inspect wing and face walls for adequate protection to personnel, erosion of slopes, loose rocks, actual or potential slides, scouring, or undermining of walls.		
Check door and gate operating and locking devices for proper operation and rust, corrosion, loose, missing, or damaged parts.		
Inspect concrete floors for cracks, breaks, scaling, and other damage.		
Check earth and gravel floors for proper grading and drainage.		
Check tracks for alignment, rails for damage and adequate connections and supports, and ties for rot or other damage.		
Inspect linings for leaks, settlement, breaks or other damage and deterioration.		
Check unlined tunnels for spalling, disintegration, loose or fallen rocks.		
Check metal roofs for rust, corrosion, adequate supports.		
Check pipeline tunnels for rust, corrosion, alignment, broken, leakage.		
Check pipeline supports and anchors for rust, corrosion, loose, missing, or broken parts.		

4.9: *(Continued)*

Item to Be Inspected	Date	Comments
Check drainage systems or facilities, particularly in ammunition tunnels, flooding, or ponding.		
Check condition of ventilation equipment for proper operation, rust, corrosion, loose, missing, or other damage to related parts.		
Check lighting systems and fixtures for operating condition, adequacy, and proper type.		
Check grounding connections for electrical continuity, loose, missing, corrosion, or other damage to the connections.		

Provide any suggestions or recommendations for follow-up activities or any other comments related to the inspection and maintenance activities.

4.10: Fence and Wall Inspection Checklist

FENCE AND WALL INSPECTION CHECKLIST

Maintenance Inspector's Name ______________________________

Description and Location of Items Being Inspected ______________________________

____

Instructions: Perform the specified inspection tasks. Insert the date when the activity is complete. Make any comments which are pertinent to future maintenance needs.

Item to Be Inspected	Date	Comments
Check post foundations and embedded pipe sleeves for cracks, settling, movement, ponding, corrosion, and damage.		
Check metal posts, rails, and other parts for rust, corrosion, looseness, damaged, and deteriorated paint.		
Check guard wires and brackets for rust, corrosion, sagging, deteriorated paint, and damage.		
Check wire fabric, holding wires, and clamps for rust, corrosion, looseness, and damage.		
Check gates for rust, corrosion, rot, insect and fungus infestation, alignment, and damage as required.		
Check wood posts, rails, and other parts for rot, fungus and insect infestation, and damage.		
Check for weeds, trash, excavations, washouts, and debris along fence line.		
Check electrical grounding of posts and fabric of fence and flexible connections at all gates.		

Provide any suggestions or recommendations for follow-up activities or any other comments related to the inspection and maintenance activities.

4.11: Pier, Wharf, Quaywall, and Bulkhead Inspection Checklist

PIER, WHARF, QUAYWALL, AND BULKHEAD INSPECTION CHECKLIST

Maintenance Inspector's Name ______________________________

Description and Location of Items Being Inspected ______________________________

____

Instructions: Perform the specified inspection and maintenance tasks. Insert the date when the activity is complete. Make any comments which are pertinent to future maintenance needs.

Item to Be Inspected	Date	Comments
Comply with all current safety precautions.		
Check horizontal and vertical alignment.		
Check for missing, broken or loose connections, obstructions and other hazardous conditions of curbings, handrails and catwalks.		
Check bollards, bits, cleats and capstans for wear, breaks, rough or sharp surfaces or edges and missing or loose bolts.		
Check deck drains and scuppers for loose, missing or broken screws, water ponding and other deficiencies.		
Check manhole covers and grating for rust, corrosion, bent or worn hinge pins and other damage.		
Inspect asphalt deck coverings for cracks, holes, and other damage.		
Check ladders and deck planking for rust, corrosion; broken, bent, or missing rungs; rot, termite or pest infestations.		
Check wood stringers, pile caps and bearing piles for missing, broken, decayed or termite and pest infestation.		
Check grounding connections for security.		

Provide any suggestions or recommendations for follow-up activities or any other comments related to the inspection and maintenance activities.

4.12: Railroad Trackage Inspection Checklist

RAILROAD TRACKAGE INSPECTION CHECKLIST

Maintenance Inspector's Name ______________________________

Description and Location of Items Being Inspected ______________________________

____

Instructions: Perform the specified inspection tasks. Insert the date when the activity is complete. Make any comments which are pertinent to future maintenance needs.

Item to Be Inspected	Date	Comments
Inspect tracks for vertical and horizontal alignment, sinking, churning, inadequate expansion; examine closely where track passes from earth fill to bridges or trestles.		
Inspect rails for breaks, splits; cracks in head, web, or base; damage from flat wheels; creeping or shoving, particularly at curves or ends; battering, flowing, chipping, slivers, and engine burn.		
Check joints for loose spikes (four spikes per tie), improper tie plate seating; improper support of rail.		
Check flangeways of firder-type rails; tops should be flush with pavement or crossing.		
Check road crossings for poor condition, roughness to road traffic, and obstructions.		
Inspect ties for decay, splitting, deterioration, rail cutting, insufficient or improper embedment in ballast, and inadequate drainage.		
Inspect ballast for dirt and mud accumulations, soft or wet spots, grass or weeds, erosion or settlement, inadequate extension beyond ties and improper slope.		
Check turnouts for lubrication, debris or dirt, inadequately spiked; operating condition of switches, switch latches, targets and lamps.		
At tank car unloading tracks, check for bonding wires across rail joints, between rails and unloading header pipe lines, and connections between rails and ground rods and insulated rail joints.		

4.12: *(Continued)*

Item to Be Inspected	Date	Comments
Check warning signs for adequacy, placement, legibility, telltales proper placed, stability of bumper blocks and cattle guards.		
Check guard rails for condition and placement.		
Check horizontal and vertical clearances at structures.		
Check for weeds, trees, or other obstructions blocking view, creating fire hazard, or reducing clearances.		

Provide any suggestions or recommendations for follow-up activities or any other comments related to the inspection and maintenance activities.

4.13: Railroad Crossing Signal Inspection Checklist

RAILROAD CROSSING SIGNAL INSPECTION CHECKLIST

Maintenance Inspector's Name ______________________________

Description and Location of Items Being Inspected ______________________________

____

Instructions: Perform the specified inspection and maintenance tasks. Insert the date when the activity is complete. Make any comments which are pertinent to future maintenance needs.

Item to Be Inspected	Date	Comments
Tighten all loose connections. Repair or replace broken wires.		
Test signal lamps by connecting a jumper across rails. Replace bulbs as needed.		
Check batteries and remove all oxidation; tighten terminal connections. Fill with water to line marker.		
Make circuit test with appropriate meter. Adjust current to trip coil relay to maximum of 70 mils and 0.5 volts.		
Check rails in operating area.		
Inspect batteries as follows: A. Each panel in this two-panel indicator is of the same size but of different thickness in order to provide a progressive indication of exhaustion. Perforation starts at left-hand panel. When this panel is completely eaten away, the cell has delivered 75% of its rated capacity. The right-hand panel then continues to perforate and the entire perforation of both panels indicates the cell has delivered rated capacity. B. With this type of progressive indicator panel, renew an entire battery when the majority of the right-hand panel shows complete perforation. (The battery has then delivered its rated capacity.)		
Inspect all painted surfaces for damage from rust and corrosion. Remove rust and corrosion and apply paint where applicable.		

Provide any suggestions or recommendations for follow-up activities or any other comments related to the inspection and maintenance activities.

4.14: Storm Drainage System Inspection Checklist

STORM DRAINAGE SYSTEM INSPECTION CHECKLIST

Maintenance Inspector's Name ____________________

Description and Location of Items Being Inspected ____________________

____

Instructions: Perform the specified inspection and maintenance tasks. Insert the date when the activity is complete. Make any comments which are pertinent to future maintenance needs.

Item to Be Inspected	Date	Comments
Comply with all current safety precautions.		
Check invert elevation of nonsedimentation basin and pipe.		
Inspect catch basins and curb inlets for debris, obstructions, and cracked, broken, or properly seated grating, settlement.		
Inspect pipelines for alignment, settlement, cracked, broken, open joints, sediment, debris, tree roots, erosion in concrete pipes, erosion and corrosion in corrugated metal pipes.		
Inspect headwalls for cracked, broken, spalling, exposed reinforcing, settlement, undermining, and condition of pipe joint at headwall.		
Check approach channels for water channeling under and around pipe or headwall.		
Inspect outfall and channel beyond headwall for sediment, debris, other obstructions, erosion of adjoining property.		
Check tide gates for motion; closure and outfall line and bar screens for sediment and obstructions.		
Check drop structures and spillways for silt and erosion.		
Inspect manhole frames and covers for rust, corrosion, fit; ladder rungs for rust, corrosion, broken parts, damaged supports.		

4.14: *(Continued)*

Item to Be Inspected	Date	Comments
Inspect manhole walls for cracking, spalling, exposed reinforcing; eroded or sandy mortar joints, loose, broken, or displaced brick.		
Check manhole bottoms for clogging, flow, silt, sewer pipe fragments, invert elevation of outlet pipe flush with bottom.		
Inspect culverts for sediment, obstructions at inlets and outlets, ditch bottoms not flush with pipe inverts, and channeling.		
Inspect gutters and ditches for cracked, broken, eroded concrete surfaces, expansion joint, alignment obstructions, ponding of water, silting or sloughing off of sides, adequate side vegetation coverage necessary to prevent erosion.		
Check for standing water which would permit mosquito breeding in drainage system.		

Provide any suggestions or recommendations for follow-up activities or any other comments related to the inspection and maintenance activities.

4.15: Refuse and Garbage Disposal Inspection Checklist

REFUSE AND GARBAGE DISPOSAL INSPECTION CHECKLIST

Maintenance Inspector's Name ______________________________

Description and Location of Items Being Inspected ______________________________

____

Instructions: Perform the specified inspection tasks. Insert the date when the activity is complete. Make any comments which are pertinent to future maintenance needs.

Item to Be Inspected	Date	Comments
Check to see that site location is less than 750 feet from public highways and human activities.		
Check to see that prevailing winds are toward habitation and highways.		
Check accessibility of fill area to trucks and equipment required in excavating, filling, backfilling, and compacting.		
Check soil quality that prevents ease of excavating and does not provide effective seal. (Sandy soil preferred.)		
Check for evidence of drainage polluting surface or subsurface water supplies.		
Check for potential washouts resulting from storm water when located in ravine.		
Check for depth of fill. It should not exceed 6 feet.		
Check for covering of clean fill over refuse. Minimum cover is 24 inches.		
Check for evidence of dust resulting from lack of vegetation from completed portions of fill.		
Check for evidence of rodents, flies, and/or other pests in area.		

Provide any suggestions or recommendations for follow-up activities or any other comments related to the inspection and maintenance activities.

C. SPECIAL FORMS

There are many maintenance-related activities in every company that must be included in the organization's overall management program. Most of these tasks are unique to the specific company. To assist management, special forms can be developed for the specific activity to enable them to control the respective maintenance function more effectively and efficiently.

Every company should evaluate its own operations to ascertain where such forms would help to make the management function more efficient. Once these activities are identified, appropriate documents can be developed. Some examples relating to housekeeping activities are included in this part of the section.

This section concludes with the presentation of several forms that can be used in a company's grounds maintenance program, including for the control of maintenance of vehicles.

4.16: Building and Grounds Work Order

BUILDING AND GROUNDS WORK ORDER

Location ______________________ Room Area ______________________

Date Prepared ______________________ Date Assigned ______________________

Craft	Description—Quantity—Material	Est. Hrs.	Act. Hrs.	Dates Worked
Carpenter				
Electrician				
Locksmith				
Painter				
Plumber				
General Maintenance				
Delivery				
Grounds				
Others				
			W/O No. ______	

Prepared By ______________________

Completed By ______________________

4.16: *(Continued)*

FORM 4.16

Key Use of Form:

This form is used to order maintenance work to be performed on building and grounds-related items. It is also used to record the actual time spent on performing the item along with the date(s).

Who Prepares:

Plant engineering personnel will complete all parts except actual date and hours worked. This will be completed by the field supervisor.

Who Uses:

Plant and engineering and maintenance department personnel.

How to Complete:

Provide the information requested in the appropriate spaces.

Alternative Forms:

A similar company-specific form can be developed by making the appropriate changes to this form.

4.17: Building and Grounds Work Schedule

BUILDING AND GROUNDS WORK SCHEDULE

Facility ____

Prepared By ______________________________ Date ______________

Time	Room or Area	Work Activity	Sq. Ft.

FORM 4.17

Key Use of Form:

To schedule, on a daily basis, the work to be performed.

Who Prepares:

The scheduling unit personnel.

Who Uses:

Plant engineering and maintenance department personnel.

How to Complete:

Place the requested information in the spaces provided.

Alternative Forms:

A company can revise the headings on this form to make it more specific for use within its organization.

4:18: Vehicle Maintenance Report

VEHICLE MAINTENANCE REPORT

Maintenance Inspector's Name ______________________ Date of Inspection ____________

Vehicle No. ____________ Year and Make ____________ Service Date ____________ Time In ________ Time Out ________

Electrical	Good	Bad	New	**Steering/Suspension**	Good	Bad	New	**Brake System**	Good	Bad	New
Battery				Tires				Fluid			
Cables				Shocks				Lines			
Alternator				Alignment				Pads			
Generator				Ball Joints				Rotors			
Regulator				Tie Rod/Ends				Springs			
Distributor				Idler Arms				Drums			
Rotor				Springs				Master Cylinder			
Condensor				P.S. Pump				Linings			
Coil				Fluid				Linkage			
Spark Plugs				Belts				Vacuum Booster			
Plug Wires				Wheel Bearings				Parking Brake			
Ignition Switch				Lube/Grease				Cables/Clamps			
Fuses											
Timing											
Headlights				**Cooling System**				**Fuel System**			
Turn Signals											
Flashers				Radiator				Carburetor			
Int. Lights				Hoses/Clamps				Air Filter			
Aux. Lights				Belts				Gas Filter			
Wipers/Blades				Antifreeze				Crankcase Filter			
				Water Pump				Fuel Pump			
				Oil Pump				Choke			
				Oil Filter				Accel. Linkage			
Trans./Drive				**Exhaust System**				**Hydraulic System**			
Grease				Manifold				Seals			
Fluid				Pipes				Fluid			
Filters				Mufflers				Hoses			
Linkage				PVC Valve				PTO Pump			
Clutch				Air Pump				Gear Box			
Drive Shaft				Heat Riser				Linkage			
Differential				Clamps				Filters			
Univ. Joint				Hangers				Rams			
Lines											

4.18: *(Continued)*

FORM 4.18

Key Use of Form:

To record results of vehicle inspection. Will also be used to plan and schedule vehicle maintenance.

Who Prepares:

Person performing the inspection.

Who Uses:

Personnel in department responsible for vehicle maintenance.

How to Complete:

Inspect all the items listed and rate the condition using one of the three choices.

Alternative Forms:

Can change form to make company-specific by making appropriate changes.

4.19: Vehicle Inspection Report

VEHICLE INSPECTION REPORT

Vehicle No.	Speedometer	
	Finish	Start

Miles		Material Hauled
From	To	

Gasoline (Check One) ☐ Full ☐ Half ☐ Quarter ☐ Empty	Oil

Are the following items in working order?

Item	Yes	No	Item	Yes	No
Foot Brake			Windshield Wipers/Washers		
Hand Brake			Gauges & Warning Indicators		
Steering			Fire Extinguisher		
Seat Belts			Flares and Flags		
Lights			Tires, Wheels & Rims		
Horn			First Aid Kit		
Rear Vision Mirrors			Spare Fuses		
Directional Signals			Chains		

Vehicle Operated By Employee		Did you and all passengers wear seat belts? ☐ Yes ☐ No
Date	Time	
Vehicle Released By Employee		Do you have any accident report blanks in the vehicle? ☐ Yes ☐ No
Date	Time	

Report Accidents Today, If Any

Report Other Mechanical Defects, If Any

Reporting Driver	Maintenance Action
Name ______________	Date ______________
Date ______________	☐ Repairs Made ☐ No Repairs Needed
Reviewing Driver	W.O. No. ______________
Name ______________	Approved by ______________
Date ______________	Location ______________

4.19: *(Continued)*

FORM 4.19

Key Use of Form:

To record the results of the inspection of a vehicle. Will also be used to plan and schedule vehicle maintenance.

Who Prepares:

Person making inspection.

Who Uses:

Personnel in department responsible for vehicle maintenance.

How to Complete:

Inspect all the items listed and provide the information requested.

Alternative Forms:

Can change form to make company-specific by making appropriate changes.

4.20: Preventive Maintenance Record for Vehicles

PREVENTIVE MAINTENANCE RECORD FOR VEHICLES

A. VEHICLE IDENTIFICATION

Make ____________ Model ____________

Year ________ Vehicle No. ________ Plate No. ________

Serial No. ____________ Date Purchased ____________

If Leased, Lessor ________________________

B. TIRE INFORMATION

Make	Size	Ply No.	Date Instl.	Mileage

Date Removed	Mileage	Comments

C. OIL AND LUBRICATION RECORD

Key to Services
A —Motor Oil Change
A1—Filter Change
B —Lubrication
B1—Gearlube
C —Wheel Bearing Service

Mileage	Services					Qts. of Oil	Job Performed	
	A	A1	B	B1	C		Date	By

4.20: *(Continued)*

D. INSPECTION AND REPAIR RECORD

Date	Mileage	Nature of Inspection, Maintenance, and/or Repair Service	Next Due Date	Cost

FORM 4.20

Key Use of Form:

To record all inspection and maintenance performed on a specific vehicle. Also used to plan and schedule future maintenance.

Who Prepares:

Personnel in the maintenance department who is responsible for caring for vehicles.

Who Uses:

Personnel who controls vehicle maintenance.

How to Complete:

Provide the information requested in the appropriate spaces in Parts A and B when first purchasing the vehicle and after the respective maintenance activities are performed.

Alternative Forms:

A company can make specific revisions to the form to meet its unique needs.

4.21: Vehicle Maintenance Schedule

VEHICLE MAINTENANCE SCHEDULE

Completed By ______________________ Date ______________

VEHICLE IDENTIFICATION	DATES MAINTENANCE IS TO BE DONE			
	Spring	Summer	Fall	Winter

4.21: *(Continued)*

FORM 4.21

Key Use of Form:

To schedule maintenance of vehicles.

Who Prepares:

Person responsible in the maintenance department who performs scheduling function.

Who Uses:

Designated maintenance personnel who orders work to be performed.

How to Complete:

Provide vehicle identification information along with the date maintenance is to be done. Place in spaces provided.

Alternative Forms:

This form can be revised to meet specific company needs.

4.22: Building Herbicide Log

BUILDING HERBICIDE LOG

Prepared By ______________________________ Date ______________

Date or Week Ending	Building	Herbicide	Amount Used	Applicator

FORM 4.22

Key Use of Form:

Documents the application of herbicides so maintenance can better control activity in addition to planning and scheduling future applications.

Who Prepares:

Designated maintenance department personnel with input from applicator.

Who Uses:

Maintenance department personnel.

How to Complete:

Provide the requested information in the appropriate spaces.

Alternative Forms:

Company can make specific revisions to the form to customize it for its specific needs.

Section 5

HOUSEKEEPING

To increase the lifetime of your facility and its contents, an effective housekeeping program is needed. To make such a program efficient, the most up-to-date forms should be used to control the activities. This section includes checklists, scheduling formats, and many other forms that can be used in the management of the housekeeping function within any organization.

A. SCHEDULE, WORK ASSIGNMENT, AND REPORT FORMS

The first series of forms in this section are used to schedule housekeeping work. These are followed by formats that are used to make the actual work assignments. Finally, forms that can be used to document the performance and completion of the maintenance work are provided.

5.1: Master Housekeeping Schedule

MASTER HOUSEKEEPING SCHEDULE

Supervisor ______________________________ Week of ______________

Work Type Abbreviations

Vac.—Vacation
S. L.—Sick Leave
Sus.—Suspension
Hol.—Holiday
Comp.—Compensation
D. O.—Day Off

A—Wax & Strip
B—Spot Wax
C—Gen. Housekeeping
D—Wet Mop
E—Wall Washing

G—Elevator Cleaning
H—Stairway Cleaning
I —Lighting
J —Utility Work
K—Other

W.O. No.	Work Type	Mon.	Tues.	Scheduled Hours			Sat.	Sun.	Tot.	Accum. Tot.
				Wed.	Thurs.	Fri.				

FORM 5.1

Key Use of Form:

To schedule housekeeping maintenance activities.

Who Prepares:

Person responsible in organization for scheduling of maintenance activities.

Who Uses:

Supervisor in charge of performing the work.

How to Complete:

The work order (W.O.) number is placed in the appropriate space along with the type of work to be performed using the noted abbreviations on top of the form. The clock hours for each day of the week are noted as is the total hours for each week. A copy of this form will be kept by the scheduler so he or she can follow up at the end of the week to evaluate whether or not the work was completed within the times noted on it.

Alternative Forms:

A company may wish to revise the form to meet their specific needs.

5.2: Building Cleaning Schedule

BUILDING CLEANING SCHEDULE

Week of: ______________________, 19 ____

NAME ▼	MONDAY						TUESDAY						WEDNESDAY						THURSDAY						FRIDAY					
	Assignments	R/E	TIMES Start	Stop	Act'l	W.O.COMP.	Assignments	R/E	TIMES Start	Stop	Act'l	W.O.COMP.	Assignments	R/E	TIMES Start	Stop	Act'l	W.O.COMP.	Assignments	R/E	TIMES Start	Stop	Act'l	W.O.COMP.	Assignments	R/E	TIMES Start	Stop	Act'l	W.O.COMP.
	TOTAL ▷						TOTAL ▷						TOTAL ▷						TOTAL ▷						TOTAL ▷					
	TOTAL ▷						TOTAL ▷						TOTAL ▷						TOTAL ▷						TOTAL ▷					
	TOTAL ▷						TOTAL ▷						TOTAL ▷						TOTAL ▷						TOTAL ▷					
	TOTAL ▷						TOTAL ▷						TOTAL ▷						TOTAL ▷						TOTAL ▷					
	GRAND TOTAL ▶						GRAND TOTAL ▶						GRAND TOTAL ▶						GRAND TOTAL ▶						GRAND TOTAL ▶					

5.2: *(Continued)*

FORM 5.2

Key Use of Form:

To schedule maintenance housekeeping (cleaning) activities.

Who Prepares:

Person responsible for scheduling maintenance activities.

Who Uses:

Used by appropriate maintenance manager to make assignments for the cleaning activities.

How to Complete:

The area designation and the items to be cleaned are placed in the appropriate spaces on the form. Once this is completed, the frequency designation corresponding to how often a specific item in a specific area is to be cleaned is placed in the respective space on the form. The "R" stands for a regular assignment and "E" an emergency one. This provides a priority for the work. W.O.Comp. notes the date the item is finished.

Alternative Forms:

Company can make revisions to the form to meet specific needs. Also, Form 5.1 can be used in place of this one.

5.3: Custodial Area Assignment

CUSTODIAL AREA ASSIGNMENT

To ____________________ Building ____________________ Time Allotted __________

General description of area to be serviced:

____

____

It is intended that each written assignment and the periodic cleaning schedule provide a full eight hours of work. If this assigned area fails to include enough work to keep busy, contact the person in charge immediately and the situation will be corrected. The assignment details the work and, under normal conditions, the time the work is to be performed. Supervision may find it necessary from time to time to adjust the assignment. If there are any questions, please discuss them with the person in charge.

Tasks or Combination of Tasks	Hours	
	From	To

Miscellaneous:

a. Perform periodic cleaning
b. Remove ice and snow from entrances
c. Keep custodial areas, equipment, and supplies neat, clean, and orderly
d. Report repairs needed
e. Turn out all unnecessary lights
f. Clean bulletin board glass and all inside glass in your area excluding exterior windows

Date ____________________ Approved By ______________________________

5.3: *(Continued)*

FORM 5.3

Key Use of Form:

Used to make specific housekeeping maintenance assignments to custodial personnel.

Who Prepares:

Completed on a daily basis by the person requesting the performance of the work or the one that schedules it.

Who Uses:

Maintenance crew supervisor.

How to Complete:

Provide the information in the appropriate spaces on the form.

Alternative Forms:

Form could be revised to meet specific needs of a company.

5.4: Facilities Project Work Report

FACILITIES PROJECT WORK REPORT

Completed By ______________________________ Date ______________

Project	Work Assigned To	Floor	Building	Time Completed
Wash & Change Ceiling Panels				
Automatic Floor Scrubber				
Strip & Refinish Floor				
Shampoo Rugs				
Shampoo Upholstered Furniture				
Service Lights				
Wash, Polish & Oil Wood Furniture				
Wash Windows				
Clean Stove & Hoods				
Move Furniture				
Other				

GENERAL NOTES:

FORM 5.4

Key Use of Form:

Daily report on assigned maintenance work and completed time.

Who Prepares:

Maintenance crew supervisor.

Who Uses:

Designated maintenance manager to evaluate efficiency of work performed.

How to Complete:

Provide the information requested in the appropriate spaces.

Alternative Forms:

Form can be revised to meet specific company needs. Other similar forms in this section can also be used if found to be appropriate.

5.5: Daily Periodic Area Cleaning Report

DAILY PERIODIC AREA CLEANING REPORT

Building ______________ Location ______________ Date __________

Custodian ______________________ Hours ____________

Area—Use
L —Lab
O —Office
R —Restroom
H —Hallway
C —Classroom

Type of Cleaning
PD —Periodic Dusting
CW—Clean-Wax
G —Glass

Room No. or Area Description	Area Use	Type of Cleaning	Number of Furniture Items					Other Info.
			Chairs	Desks	Tables	Files	Bookcases	

Supervisor's Name: ______________________

FORM 5.5

Key Use of Form:

Assign work and note when completed by supervisor approval.

Who Prepares:

Maintenance supervisor.

Who Uses:

Personnel designated to receive the report and use to record the completion of the work and evaluate efficiency.

How to Complete:

Provide the information requested in the appropriate spaces.

Alternative Forms:

Can make changes to the form to meet specific company needs.

5.6: Daily Floor Maintenance Work Report

DAILY FLOOR MAINTENANCE WORK REPORT

Building ____________ Location ____________ Date ________

Crew names and amount of hours each worked:

Area—Use

L —Lab
O —Office
R —Restroom
H —Hallway
C —Classroom

Procedure

C —Clean
W —2 Coats Wax
SS —Safe-T-San®
T —Terrazzine®
D —Dura Seal®
R —Renovate
N —Numastic®
S —Strip
F —2 Coats Fin.
SE —Seal

Room No. or Area Description	Area Use	Floor Type	Procedure Used	Sq. Ft.	Other Info.

Supervisor's Name: ____________

FORM 5.6

Key Use of Form:

Assign work and note when completed by supervisor approval.

Who Prepares:

Maintenance supervisor.

Who Uses:

Personnel designated to receive the report and use to record the completion of the work and evaluate efficiency.

How to Complete:

Provide the information requested in the appropriate spaces.

Alternative Forms:

Can make changes to the form to meet specific company needs.

5.7: Daily Window Washing Work Report

DAILY WINDOW WASHING WORK REPORT

Building ______________________ Location ______________________ Date __________

Crew names and amount of hours each worked:

Room No. or Area Description	Venetian Blinds				Glass Washed			Other Info.
	Remove	Wash	Install	Dust	Windows	Storm Windows	Screens	

Supervisor's Name: ______________________________

FORM 5.7

Key Use of Form:

Assign work and note when completed by supervisor approval.

Who Prepares:

Maintenance supervisor.

Who Uses:

Personnel designated to receive the report and use to record the completion of the work and evaluate efficiency.

How to Complete:

Provide the information requested in the appropriate spaces.

Alternative Forms:

Can make changes to the form to meet specific company needs.

5.8: Carpet and Upholstery Work Report

CARPET AND UPHOLSTERY WORK REPORT

Building ______________________ Location ______________________ Date ___________

Crew names and amount of hours each worked:

Area—Use

L—Lab
O—Office
C—Classroom
H—Hallway
ST—Stairs
R —Reception
E —Entrance

Procedure

PL —Pile Lift
SP —Spot Clean
DF —Dry Foam Shampoo
AS —Anti-Static Treatment
OC—Odor Control Treatment

Room No. or Area Description	Area Use	Type	Color	Procedure	Amount and Type of Cleaning Materials	Sq. Ft.

Supervisor's Name: ______________________________

FORM 5.8

Key Use of Form:

Assign work and note when completed by supervisor approval.

Who Prepares:

Maintenance supervisor.

Who Uses:

Personnel designated to receive the report and use to record the completion of the work and evaluate efficiency.

How to Complete:

Provide the information requested in the appropriate spaces.

Alternative Forms:

Can make changes to the form to meet specific company needs.

B. INSPECTION FORMS

This section contains a variety of forms used to inspect the condition of items cared for by the housekeeping unit of the maintenance organization. The information on the forms is used to evaluate the housekeeping activity. It is also used to plan and schedule future similar activities. Finally, the information is placed on the appropriate inventory form so the designated manager can better control the maintenance of the respective item.

5.9: Room Housekeeping Checklist

ROOM HOUSEKEEPING CHECKLIST

Completed By ______________________________ Date ______________

Building ______________________ Location ______________________

Room Designation	Time Serviced	Materials Used	Other Services Required

GENERAL COMMENTS:

FORM 5.9

Key Use of Form:

To record housekeeping work completed, the time it took, and materials used. Also used to evaluate performance of maintenance crew.

Who Prepares:

Maintenance supervisor.

Who Uses:

Designated maintenance manager who evaluates efficiency of crew.

How to Complete:

Provide the information requested in the appropriate spaces.

Alternative Forms:

Can make revisions to this form to meet specific equipment needs.

5.10: Custodial Area Sanitation Report

CUSTODIAL AREA SANITATION REPORT

Maintenance Inspector's Name ______________________________

Building ______________________ Date of Inspection ______________

DI —Dirty
DU—Dusty
S —Smeared/Stained
L —Littered

WX—Needs Wax
CW—Cobwebs
O —Organization
R —Repair Needed

WA—Needs Washing
SV —Service Needed
/ —Acceptable

Item	Room No. or Area	Key	Item	Room No. or Area	Key
Mirrors			Door Glass		
Shelves			Transoms		
Soap Display			Blackboards		
Sinks			Chalktrays		
Towel Cabinets			Erasers		
Supplies			Tables		
Sanitary Cabinets			Book Cases		
Urinals			Desks		
Toilet Bowls			Desk Lamps		
Toilet Paper			Ash Trays		
Tile			Telephones		
Furniture			Chairs		
Equipment			File Cabinets		
Metal Partition			Smoking Units		
Waste Cans			Steps/Stairs		
Floors			Handrails		
Radiators			Drinking Fountains		
Window Ledges			Bulletin Boards		
Wood Trim			Pipes/Ducts		
Baseboards			Rest Room Doors		

GENERAL COMMENTS:

5.10: *(Continued)*

FORM 5.10

Key Use of Form:

To record the results of inspecting the completion and cleanliness of items on the form.

Who Prepares:

Maintenance inspector.

Who Uses:

Designated maintenance manager who uses it to evaluate efficiency of work and to schedule and order any necessary follow-up maintenance and repairs.

How to Complete:

Complete the form by providing the information requested in the space. Also, using the key, evaluate the cleanliness of each item.

Alternative Forms:

The form can be revised to meet specific company needs.

5.11: Custodial Inspection Report—Floor Maintenance

CUSTODIAL INSPECTION REPORT—FLOOR MAINTENANCE

Date of Inspection ________ Crew Inspected (Names) ________________________

Building ______________ Location ______________ Areas Inspected ______________

Floor Type __________ Floor Condition ☐ Good ☐ Fair ☐ Poor

Condition of Floor Finish ☐ Good ☐ Fair ☐ Poor

I. PROCEDURE USED

Dust mopping	☐
Dry cleaning	☐
Cleaning w/finish	☐
Renovating	☐
Water mopping	☐
Machine scrub	☐
Auto scrub	☐
Water pickup	☐
Mop/bucket	☐
Water vac	☐
Rinsing	☐
Seal: 1 coat	☐
2 coats	☐
3 coats	☐
Brand __________	
Finish: 1 coat	☐
2 coats	☐
3 coats	☐
Brand __________	
Cleaner Used:	
Brand __________	

II. EQUIPMENT CONDITION

	Clean	Needs Replacement
Dry mops	☐	☐
Wet mops	☐	☐
Mop units	☐	☐
Floor machine	☐	☐

III. WORK EVALUATION

1—Unsatisfactory 2—Marginal
3—Average 4—Above Average
5—Outstanding

Quality of Finish Floor 1 2 3 4 5
(circle one)

Streaking	☐ Yes	☐ No
Caused by	☐ Products	☐ Employees
Splash marks	☐ Yes	☐ No
Lap marks	☐ Yes	☐ No
Clean baseboards	☐ Yes	☐ No
Void spots	☐ Yes	☐ No
Build-up	☐ Yes	☐ No
Powdering	☐ Yes	☐ No

IV. PERSONNEL EVALUATION—Evaluate each crew member annually using another form.

V. SUMMARY

I Approve/Disapprove of this evaluation. (Circle one)

Supervisor ____

5.11: *(Continued)*

FORM 5.11

Key Use of Form:

Evaluate the performance of the housekeeping staff.

Who Prepares:

Maintenance inspector.

Who Uses:

Designated maintenance management to review efficiency of work.

How to Complete:

Mark the appropriate boxes as the inspection is made. Also, provide other information requested in the appropriate spaces. The supervisor of the crew is also provided an opportunity to agree or disagree with the evaluation.

Alternative Forms:

Revisions can be made to the form to make it company-specific.

5.12: Custodial Inspection Report—Window Maintenance

CUSTODIAL INSPECTION REPORT—WINDOW MAINTENANCE

Date of Inspection ____________ Crew Inspected (Names) ____________________

Building ______________ Location ______________ Area Inspected ______________

I. WINDOW WASHING
(Condition of windows: Heavy Soil ☐ Average Soil ☐ Light Soil ☐)

A. Procedure Used		B. Equipment		C. Supplies	
Belts	☐	Buckets	☐	Cleaner used:	
Window Pole	☐	Sponges	☐	Ammonia	☐
____________	☐	Cheesecloth	☐	________________	☐
____________	☐	Squeegee	☐	________________	☐
____________	☐	____________	☐	________________	☐

D. Condition of Supplies and Equipment

	Adequate	Inadequate	Dirty	Clean	Needs Replacement
Equipment	☐	☐	☐	☐	☐
Supplies	☐	☐	☐	☐	☐
Belts	☐	☐	☐	☐	☐
Ladders	☐	☐	☐	☐	☐

II. DUSTING
(Condition of venetian blinds: Heavy Soil ☐ Average Soil ☐ Light Soil ☐)

A. Procedure Used		B. Equipment		C. Supplies	
Vacuuming	☐	Back vac	☐	________________	☐
Washing	☐	Vacuum	☐	________________	☐
____________	☐	Duster	☐	________________	☐

D. Condition of Equipment and Supplies

	Adequate	Inadequate	Dirty	Clean	Needs Replacement
Vacuum	☐	☐	☐	☐	☐
__________	☐	☐	☐	☐	☐
__________	☐	☐	☐	☐	☐

III. WORK EVALUATION
(1—Unsatisfactory, 2—Marginal, 3—Average, 4—Above Average, 5—Outstanding)

A. Quality of Finished Work 1 2 3 4 5 (Circle one)

Streaky windows	☐ Yes	☐ No
Blinds dusty	☐ Adequate	☐ Inadequate
______________	☐ Accidents	☐ No accidents

B. Individual Evaluation—Evaluate each crew member annually using another form.

5.12: *(Continued)*

IV. VERBAL SUMMARY

I Approve/Disapprove of this evaluation. (Circle one)

Supervisor ________________________________

FORM 5.12

Key Use of Form:

Evaluate the performance of the housekeeping staff.

Who Prepares:

Maintenance inspector.

Who Uses:

Designated maintenance management to review efficiency of work.

How to Complete:

Mark the appropriate boxes as the inspection is made. Also, provide other information requested in the appropriate spaces. The supervisor of the crew is also provided an opportunity to agree or disagree with the evaluation.

Alternative Forms:

Revisions can be made to the form to make it company-specific.

5.13: Housekeeping Inspection Form

HOUSEKEEPING INSPECTION FORM

Facility ____________________ Location ____________________

Maintenance Inspector's Name ____________________ Date of Inspection ____________

AREA	RATING			
	Poor	Fair	Good	Excellent
ENTRANCE: Carpet				
Glass, metal surfaces				
Corners				
Floor				
LOBBIES: Dusting				
Floor appearance				
Sweeping, vacuuming				
Spot cleaning				
Fixtures				
Water fountains				
ELEVATORS: Treads				
Lights				
Walls, doors				
Floors, carpet				
CORRIDORS: Sweeping, vacuuming				
Floor appearance				
Baseboards				
Spot cleaning				
Water fountains				
STAIRWELLS: Rails, walls				
Steps, landings				

5.13 *(Continued)*

AREA	RATING			
	Poor	Fair	Good	Excellent
RESTROOMS: Dispensers				
Basins				
Floors				
Mirrors				
Partitions				
Toilets, urinals				
Waste cans				
Walls, doors				
OFFICE EQUIPMENT AREAS: Ash trays				
Furniture equipment				
Door kick plates				
Phones, lamps				
Walls, doors, spot cleaning				
Waste baskets				
Partitions				
Low dusting				
Floor appearance				
Sweeping, vacuuming				
Baseboards				
Corners				
WINDOWS: Glass				
Sills, frames				
Blinds				
EXTERIOR AND GROUNDS: Policing				
Sidewalks				
Entrance area				

5.13: *(Continued)*

AREA	RATING			
	Poor	Fair	Good	Excellent
EXTERIOR AND GROUNDS (*continued*): Stairwell, equipment drop				
Lawn area				
Parking area				
Trash area				
JANITOR CLOSETS: Cleanliness, organization				
Supplies, equipment				
MISCELLANEOUS:				
Air vents				
Exit lights				
Carpet spotting				
Cafeteria				
Phone booths				
Storage areas				
Other				

COMMENTS:

FORM 5.13

Key Use of Form:

Used to record the results of the inspection.

Who Prepares:

Maintenance inspector.

Who Uses:

Designated maintenance management personnel for analysis of work performed and to plan and schedule any required follow-up work.

How to Complete:

Rate the work performed by marking the appropriate spaces provided.

Alternative Forms:

A company can make appropriate revisions to the form to meet specific needs.

5.14: Supervisor's Checklist and Report

SUPERVISOR'S CHECKLIST AND REPORT

Building ______________________ Location ______________________

Date of Inspection ______________________ Janitor ______________________

Item	Good	Fair	Poor
Floors—Look in corners and behind doors			
Furniture—Desks, chairs, tables			
Radiators—Look underneath			
Baseboard moldings			
Chair rails			
Window sills			
Partitions—Look at sides and tops			
Supply closets			
Slop sink areas			
Wall spotting			
Stairwells, landings, lobbies, hallways			
Desk mats			
Toilet Room Floors			
Bowls—Look at base			
Seats—Top and bottom			
Towel dispenser—Look on top			
Soap dispensers			
Sanitary napkin dispensers			
Tiled walls			
Mirrors			
Metal Work			
Partitions—Tops			
All disposal cans			
Water coolers			
Telephones			

5.14: *(Continued)*

Item	Good	Fair	Poor
Overhead Blinds			
Light fixtures			
Picture frames			
Other Items:			

GENERAL COMMENTS:

Supervisor's Signature ______________________

FORM 5.14

Key Use of Form:

Used to record the results of the inspection.

Who Prepares:

Maintenance inspector.

Who Uses:

Designated maintenance management personnel for analysis of work performed and to plan and schedule any required follow-up work.

How to Complete:

Rate the work performed by marking the appropriate spaces provided.

Alternative Forms:

A company can make appropriate revisions to the form to meet specific needs.

5.15: Evaluation of the Custodial and Maintenance Program for School Buildings

EVALUATION OF THE CUSTODIAL AND MAINTENANCE PROGRAM FOR SCHOOL BUILDINGS

Location ______________________ Date of Inspection ______________________

Maintenance Inspector's Name ______________________________________

Rating System: Column No. 1—Superior = 4 points
Column No. 2—Above Average = 3 points
Column No. 3—Average = 2 points
Column No. 4—Poor = 1 point

Items	No. 1	No. 2	No. 3	No. 4	Remarks
CUSTODIANS: Dress					
Attitude					
Performance					
Capability					
GROUNDS: Condition of lawn					
Condition of shrubbery					
Condition of drives					
Condition of walks					
Cleanliness of grounds					
Drainage					
Condition of fences					
Playground equipment					
Condition of flag					
OUTSIDE OF BUILDING: Gravel stop					
Roof					
Outside trim					
Gutters and downspout					
Glazing of windows					
Foundations					
Structure					

5.15: *(Continued)*

Items	No. 1	No. 2	No. 3	No. 4	Remarks
CORRIDORS: Outside doors					
Floor mats					
Condition of floors					
Condition of walls					
Condition of skylights					
Cleanliness of light fixtures					
Condition of ceiling					
Drinking fountains					
Lockers and shelves					
Panic hardware					
ADMINISTRATIVE UNIT: Condition of floors					
Condition of walls					
Light fixtures					
Blinds and shades					
Furniture and cabinets					
Lounge Condition					
Clinic Condition					
CAFETERIA AND KITCHEN: Condition of floors					
Condition of walls					
Condition of windows					
Light fixtures					
Condition of furniture					
Condition of tables					
Cleanliness of equipment					
Screens and doors					
Cleanliness of mop sink					
Condition of storage room					
Condition of restrooms					
Condition of grease trap					

5.15: *(Continued)*

Items	No. 1	No. 2	No. 3	No. 4	Remarks
CLASSROOMS: Condition of floors					
Condition of walls					
Condition of windows					
Condition of doors					
Condition of ceiling					
Condition of light bulbs					
Cleanliness of light fixtures					
Condition of blinds and shades					
Regulation of heat					
Chalkboard condition					
Chalk rails condition					
Condition of erasers					
Furniture condition					
Convector/radiators					
Other types of heat					
Teachers' cabinets					
Shelving condition					
Toilets					
Sink condition					
Work counters					
SPECIAL INSTRUCTIONAL AREAS: *Library* Condition of workroom					
Condition of lighting					
Condition of furniture					
Shelving					
Labs Supply room condition					
Condition of sinks					

5.15: *(Continued)*

Items	No. 1	No. 2	No. 3	No. 4	Remarks
Labs (continued) Cleanliness of tables					
General appearance					
Shops Condition of floors					
Condition of walls					
Cleanliness of light fixtures					
Condition of sinks					
Work table condition					
Machinery condition					
General appearance					
TOILET ROOMS: Condition of doors					
Condition of floors					
Condition of stalls					
Supply dispenser condition					
Amount of supplies					
Cleanliness of light fixtures					
Condition of bowls					
Condition of sinks					
Condition of urinals					
General appearance					
GYMNASIUM: Condition of floor					
Lighting condition					
Seating condition					
Dressing room condition					
Toilet room condition					
Shower stall condition					
Condition of walls					
General appearance					

5.15: *(Continued)*

Items	No. 1	No. 2	No. 3	No. 4	Remarks
HEATING PLANT AND ELECTRICAL SYSTEM: Boiler room					
Grate condition					
Condition of flues					
Condition of boiler					
Condition of toilets					
Fuse panel condition					
Circuit breaker condition					
Circuits					
Switches and covers					
General condition					
CARE OF SUPPLIES AND EQUIPMENT: Storage room condition					
Condition of sinks					
Care of tools					
Tool storage					
Supply storage					
Use of supplies					
Kind of supplies					
SAFETY: No. of fire extinguishers					
Condition of fire extinguishers					
Exits open					
Corridor obstruction					
Exit lights					
Fire hazards					
Fire alarm system					
Fire escape					

5.15: *(Continued)*

Items	No. 1	No. 2	No. 3	No. 4	Remarks
GENERAL: General appearance of building					
Painting schedule					

SCORING PROCEDURE: Add up the individual columns. Sum the four columns. Divide this number by the total number of applicable items above times 4. The resulting value is the percent grade.

FORM 5.15

Key Use of Form:

Used to record the results of the inspection.

Who Prepares:

Maintenance inspector.

Who Uses:

Designated maintenance management personnel for analysis of work performed and to plan and schedule any required follow-up work.

How to Complete:

Rate the work performed by marking the appropriate spaces provided.

Alternative Forms:

A company can make appropriate revisions to the form to meet specific needs.

5.16: Exterior Maintenance Quality Assurance Evaluation

EXTERIOR MAINTENANCE QUALITY ASSURANCE EVALUATION

Building ______________________ Date of Inspection ______________

Maintenance Inspector's Name ______________________________

Rating System: 9–10 = Excellent 7–8 = Very Good 5–6 = Satisfactory 3–4 = Poor 1–2 = Unsatisfactory

10–9	8–7	6–5	4–3	2–1	Item	Remarks
LAWNS—SHRUB BEDS						
					Overall appearance	
					Trash and clutter	
					Edging—walks and beds	
					Trimming	
					Weeds	
LAWNS—ROUGH—DITCHES						
					Trash	
					Excessive weeds	
					Overall appearance	
BUILDING EXTERIOR						
					Door and window frames	
					Fascia and soffit	
					General appearance	
					Equipment	
					Roof condition	
					Signs	
					Outbuildings	
PARKING AREAS AND DRIVEWAYS						
					General appearance	
					Trash and weeds	
					Curb and gutter	
					Traffic lines	
					Crack filling	
					Truckwell	
					Signs	

5.16: *(Continued)*

10–9	8–7	6–5	4–3	2–1	Item	Remarks
CAR AND TRUCK PORTS						
					Overall appearance	
					Clutter	
					Spot painting required	
					Gas islands	
EXTERIOR STORAGE						
					Trash and weeds	
					Storage bin condition	
					Adequate trash storage	
					Oil leaks	
					Proper storage	
					Proper trailer parking	
FENCE AND GATES						
					General condition	
					Trash and weed control	
					Clearance—fence to equipment	
					Evidence of rusting	
					Gates—operation	

GENERAL NOTES:

FORM 5.16

Key Use of Form:

Used to record the results of the inspection.

Who Prepares:

Maintenance inspector.

Who Uses:

Designated maintenance management personnel for analysis of work performed and to plan and schedule any required follow-up work.

How to Complete:

Rate the work performed by marking the appropriate spaces provided.

Alternative Forms:

A company can make appropriate revisions to the form to meet specific needs.

5.17: Interior Maintenance Quality Assurance Evaluation

INTERIOR MAINTENANCE QUALITY ASSURANCE EVALUATION

Building ______________________ Date of Inspection ______________

Maintenance Inspector's Name ______________________

Rating System: 9–10 = Excellent 7–8 = Very Good 5–6 = Satisfactory
3–4 = Poor 1–2 = Unsatisfactory

10-9	8-7	6-5	4-3	2-1	Item	Remarks
RESTROOMS						
					Stools	
					Urinals	
					Washbowls	
					Partitions	
					Dispensers—soap/towel	
					Mirrors	
					Glass	
					Dusting	
					Spot cleaning—walls, benches	
					Light fixtures	
					Floor condition	
OFFICE AREA						
					Dusting—furniture, etc.	
					Spot cleaning	
					Glass	
					Floors	
					Carpet	
					Phones	
					Venetian blinds	
					Light fixtures	
					Drinking fountains	
					Stairways	
					Vending machines	

5.17: *(Continued)*

10–9	8–7	6–5	4–3	2–1	Item	Remarks
CLOSETS						
					Floors	
					Equipment	
					Labels on containers	
					Organization and neatness	
					Dusting	

GENERAL NOTES:

FORM 5.17

Key Use of Form:

Used to record the results of the inspection.

Who Prepares:

Maintenance inspector.

Who Uses:

Designated maintenance management personnel for analysis of work performed and to plan and schedule any required follow-up work.

How to Complete:

Rate the work performed by marking the appropriate spaces provided.

Alternative Forms:

A company can make appropriate revisions to the form to meet specific needs.

5.18: Shop Areas Maintenance Quality Assurance Evaluation

SHOP AREAS MAINTENANCE QUALITY ASSURANCE EVALUATION

Building ______________________ Date of Inspection ______________

Maintenance Inspector's Name ______________________________

Rating System: 9–10 = Excellent 7–8 = Very Good 5–6 = Satisfactory
3–4 = Poor 1–2 = Unsatisfactory

10–9	8–7	6–5	4–3	2–1	Item	Remarks
STOREROOMS						
					Clutter—proper storage	
					Floor condition	
					Light fixtures	
					Condition of walls	
EQUIPMENT REPAIR AREA						
					Clutter—proper storage	
					Condition of walls	
					Condition of doors	
					Condition of floors	
					Condition of wash pit	
					Condition of wash racks	
					Light fixtures	
LOADING AREAS						
					Clutter	
					Floor condition	
					Light fixtures	
					Condition of doors	
DISTRIBUTION SHOPS						
					Clutter	
					Shelving and storage	
					Light fixtures	
					Walls	
					Floor condition	
BUILDING MAINTENANCE SHOP						
					Clutter	
					Floor condition	

5.18: *(Continued)*

10–9	8–7	6–5	4–3	2–1	Item	Remarks
BUILDING MAINTENANCE SHOP *(Continued)*						
					Wall condition	
					Proper storage	
BOILER ROOM						
					Clutter	
					Equipment condition	
					Floor condition	
					Pipe covering	
					Pipe markings	
					Leaks	
					Louvers, grills, etc.	
MECHANICAL ROOMS						
					Clutter	
					Duct work	
					Piping	
					Leaks	
					Wall condition	
					Ceiling condition	
					Equipment condition	
SPECIAL GENERAL NOTES						
					Temperature	
					Lighting	
					Flammable liquid storage	
					Appropriate containers	
					Floor markings	
					Fire doors	
					Painting condition	
					Other	

GENERAL NOTES:

5.18: *(Continued)*

FORM 5.18

Key Use of Form:

Used to record the results of the inspection.

Who Prepares:

Maintenance inspector.

Who Uses:

Designated maintenance management personnel for analysis of work performed and to plan and schedule any required follow-up work.

How to Complete:

Rate the work performed by marking the appropriate spaces provided.

Alternative Forms:

A company can make appropriate revisions to the form to meet specific needs.

C. SPECIAL FORMS

There are many maintenance-related activities in every company that must be included in the organization's overall management program. Most of these tasks are unique to the specific company. To assist management, special forms can be developed for the specific activity to enable them to control the respective maintenance function more effectively and efficiently.

Every company should evaluate its own operations to ascertain where such forms would help to make the management function more efficient. Once these activities are identified, appropriate documents can be developed. Some examples relating to housekeeping activities are included in this part of the section.

5.19: Snow Removal Call-in Sheet

SNOW REMOVAL CALL-IN SHEET

Report To ____________________ Date __________ Completed By ____________________

Personnel Called	Time Called	Time Signed In	Time Signed Out

GENERAL NOTES:

FORM 5.19

Key Use of Form:

To record the results of calling in maintenance personnel to remove snow.

Who Prepares:

Person making the calls, usually the immediate supervisor or a person in charge of time keeping.

Who Uses:

Designated personnel who keeps track of this information for such things as completing payroll.

How to Complete:

Provide the information requested in the appropriate spaces.

Alternative Forms:

A company can customize the form by making appropriate changes.

5.20: Building Pesticide Log

BUILDING PESTICIDE LOG

Prepared By ______________________ Date ____________

Date or Week Ending	Building	Pesticide	Amount Used	Applicator

GENERAL NOTES:

FORM 5.20

Key Use of Form:

Keep track of the application of pesticides. Local, state and/or federal agencies may require companies to keep such records.

Who Prepares:

Designated maintenance department or plant engineering personnel.

Who Uses:

Maintenance personnel to contract activity and to plan and schedule future applications.

How to Complete:

Provide the information requested in the appropriate spaces.

Alternative Forms:

Make revisions to existing form as needed to meet specific company needs.

5.21: Building Insecticide Log

BUILDING INSECTICIDE LOG

Prepared By ____________________ Date ____________

Date or Week Ending	Building	Insecticide	Amount Used	Applicator

GENERAL NOTES:

FORM 5.21

Key Use of Form:

Keep track of the application of insecticides. Local, state and/or federal agencies may require companies to keep such records.

Who Prepares:

Designated maintenance department or plant engineering personnel.

Who Uses:

Maintenance personnel to contract activity and to plan and schedule future applications.

How to Complete:

Provide the information requested in the appropriate spaces.

Alternative Forms:

Make revisions to existing form as needed to meet specific company needs.

5.22: Supervisor's Daily Personnel Checklist

SUPERVISOR'S DAILY PERSONNEL CHECKLIST

Completed By ______________________ Date ____________

Name	Attendance	Work Quality	Work Attitude	Supplies Used	Notes

FORM 5.22

Key Use of Form:

To record the attendance, work quality, and attitude of each maintenance worker along with the supplies he/she uses in performing the maintenance activity.

Who Prepares:

Supervisor or crew.

Who Uses:

Designated maintenance supervisor and other managers to evaluate efficiency and effectiveness of each person..

How to Complete:

Provide the information requested in the appropriate spaces. The supervisor completing the form should use his/her own rating system as needed to complete the form.

Alternative Forms:

Revisions can be made to this form to make it company-specific.

Section 6

MECHANICAL SYSTEMS AND EQUIPMENT

This section includes forms used for the control of the inspection, maintenance and repair of pumps, valves, hydraulic and pneumatic instrumentation, piping, heaters, air conditioners and other mechanical equipment and systems. The maintenance and repair, and thus the management of these activities, for most structural and finish building materials are a relatively straightforward matter since any one item is comprised of essentially the same material. The subject of maintaining mechanical systems and equipment, however, is unique in that they are comprised of many different materials. Furthermore, each company that manufactures the systems and equipment has its own standards, which require a separate set of maintenance and repair procedures. Because of this, it becomes important that a company dedicate resources to develop a complete inventory of its mechanical systems and equipment in a well-documented format. In addition, because of the relative complexity of each piece of equipment, inspections must be performed in accordance with manufacturer's recommendations on a regular basis. The results of each inspection must be documented to provide information for future action.

Refer also to Section 8 for additional forms to assist in managing the maintenance function as it relates to mechanical equipment and systems.

A. INVENTORY FORMS

The first series of formats, titled record or data forms, are to be used by the management of the maintenance department to establish an inventory of all the mechanical equipment and distribution systems within the company. They are completed the same as other inventory forms. As the various pieces of equipment and systems are inspected and/or maintained, the results of same can be documented on the inventory forms. These formats should be referred to as the first step in developing a plan in solving maintenance problems. As noted earlier, computer software can be used to establish a complete inventory system.

6.1: Equipment Record

EQUIPMENT RECORD

BUILDING	LOCATION	PAGE NO.

FAN	Size	R.P.M.	Bearings	Filters—Quantity	Belts—Quantity	
Mfg.	C.F.M.	Shaft	Sheave	Size	Size	
MOTOR	H.P.	F.L. Amps	Model	Frame	Shaft	Bearings
Mfg.	Voltage	R.P.M.	Serial	Service Factor	Sheave	Temp.
PUMP	Size	Model	Bearing	Head	Coupling	
Mfg.	G.P.M.	Serial	Seal/Packing	Shaft		
REMARKS						

Service Record

Date	Service Performed	Date	Service Performed

FORM 6.1

Key Use of Form:

To inventory mechanical equipment.

Who Prepares:

Plant engineering personnel.

Who Uses:

Plant engineer or maintenance department personnel.

How to Complete:

Provide the applicable information requested in the appropriate spaces. Leave any space blank that doesn't apply to the specific piece of equipment being inventoried.

Alternative Forms:

A company may desire to revise the form to make it more specific to meet its needs. Also, other inventory forms in this section may apply.

6.2: Equipment Data Card

EQUIPMENT DATA CARD

Building		Completed By
Location		Date
Item		Req. No.
Description		
Vendor		
Manufacturer		
Estimated Life		
Project No.	Purchase Order No.	Date
Size	Capacity	Type/Model
R.P.M.	Serial No.	Weight
Packing Specifications		
Bearing Data		
Lubrication Data		
Drive Specifications		
Auxiliary Equipment Nos.		
This machine is an Auxiliary of No.		
Frequency of Inspection	By	
Items for Inspection		

	SPARE PARTS STOCK LIST			
Mfg. Part No.		**Max.**	**Min.**	**Unit Cost**

6.2: *(Continued)*

FORM 6.2

Key Use of Form:

To inventory mechanical equipment.

Who Prepares:

Plant engineering personnel.

Who Uses:

Plant engineer or maintenance department personnel.

How to Complete:

Provide the applicable information requested in the appropriate spaces. Leave any space blank that doesn't apply to the specific piece of equipment being inventoried.

Alternative Forms:

A company may desire to revise the form to make it more specific to meet its needs. Also, other inventory forms in this section may apply.

6.3: Pump Data Sheet

PUMP DATA SHEET

Prepared By ______________________________ Date ________________________________

Pump Manufacturer		No. Units Required
Pump Type		GPM
Total Discharge Head		Net Positive Suction Head
Liquid Being Pumped		
Specific Gravity	PH	TEMP
Type of Seals		
Type of Lubrication		
Application		
Special Requirements		
Motor Manufacturer		
RPM	HP	FR
Volt	Phase	HERTZ
Motor Enclosure		NEMA Design
Remarks:		

FORM 6.3

Key Use of Form:

To inventory mechanical equipment.

Who Prepares:

Plant engineering personnel.

Who Uses:

Plant engineer or maintenance department personnel.

How to Complete:

Provide the applicable information requested in the appropriate spaces. Leave any space blank that doesn't apply to the specific piece of equipment being inventoried.

Alternative Forms:

A company may desire to revise the form to make it more specific to meet its needs. Also, other inventory forms in this section may apply.

6.4: Heat Exchanger Data Sheet

HEAT EXCHANGER DATA SHEET

Prepared By ______________________ Date ______________________

Heat Exchanger Manufacturer	
Heat Exchanger Type	
Application	
Liquid Being Cooled:	Temperature In
Temperature Out	GPM
Cooling Liquid:	Temperature In
Temperature Out	GPM
Heat Exchanged	
Remarks	

FORM 6.4

Key Use of Form:

To inventory mechanical equipment.

Who Prepares:

Plant engineering personnel.

Who Uses:

Plant engineer or maintenance department personnel.

How to Complete:

Provide the applicable information requested in the appropriate spaces. Leave any space blank that doesn't apply to the specific piece of equipment being inventoried.

Alternative Forms:

A company may desire to revise the form to make it more specific to meet its needs. Also, other inventory forms in this section may apply.

B. INSPECTION FORMS

To assist management in making inspections or performing routine service on various types of mechanical equipment, appropriate checklists and survey forms are provided.

The person making the inspection should first record all the general types of information requested such as Date of Inspection, Item Being Inspected, and so on. The next step is the inspection and recording of observations on this document, along with any suggested follow-up activities that should take place such as an inspection, a need for immediate repair, or a recommendation for replacement. More than one day may be needed to complete the inspection process.

After the form is completed, it should be transmitted to the designated maintenance personnel. In addition, copies should be kept on file for future reference.

6.5: Heating, Ventilation, and Air-Conditioning Drawing Checklist

HEATING, VENTILATION, AND AIR-CONDITIONING DRAWING CHECKLIST

Project ______________________ Location ______________________

Review Performed By ______________________ Date ______________

Architectural Each heating, ventilating, and air-conditioning drawing of a project should be closely checked against architectural drawings and field conditions to assure:

____ 1. Agreement on arrangements and locations of:
- a. Columns and their coordinates
- b. Walls and partitions
- c. Windows and doors
- d. Stairs, ladders, and ramps
- e. Tunnels, pits, and areaways
- f. Names of rooms and areas, and of streets on ground floor
- g. Masonry openings (other than heating, ventilating, and air-conditioning openings)
- h. Louvers (wall and window type)
- i. Louvers (door and partition type, also undercut doors)
- j. Roof openings and scuttles
- k. Fuel tanks, manholes, vents, and fill boxes
- l. Chimneys, thimbles, and roof sleeves (and their sizes)
- m. Radiators and window air-conditioning units (and their riser furrings)
- n. Furring and shafts for ducts and pipe risers
- o. Recesses (and insulation) for convectors, unit heaters, etc.
- p. Radiators, convectors in respect to sill height
- q. Cleanout doors for chimney bases, kitchen ducts, etc.
- r. Holes in floors and walls to suit owners' equipment needs

____ 2. Provision of necessary:
- a. Access doors at duct and pipe shafts, hung ceilings, partitions, etc.
- b. Floor-to-ceiling heights in boiler rooms, equipment rooms, fan rooms
- c. Partition and door arrangements for equipment rooms
- d. Doors, corridors, and knock-out walls (and their sizes) for bringing in and taking out equipment
- e. Trenches and tunnels for pipes and ducts, and pits for equipment
- f. Concrete pads under equipment and curbs for floor penetrations by ducts
- g. Grouting around fire-damper sleeves, floors, and walls
- h. Gratings for interiors of shafts, floors, or roof openings
- i. Details of dimensions (roof, floor, wall, and partition openings including outdoor-air intakes, exhaust discharges, and combustion-air intakes)
- j. Concrete pads and curbs at floor penetrations and at doors to upper-story equipment rooms with waterproof-membrane floors

____ 3. Clearance of:
- a. Owners' inserts and special markers and equipment
- b. Acoustical panels (symmetrical penetration of ceiling by air outlets)
- c. Hung ceilings by piping and top registers

6.5: *(Continued)*

d. Furred walls and partitions
e. Windows, doors, and door swings
f. Owners' equipment holes in floor and walls, traffic aisles
g. Stairs, landings, elevator shafts, and future knock-out walls

_____ 4. Agreement of architectural construction details with heating, ventilating, and air-conditioning requirements

NOTES:

Structural Each heating, ventilating, and air-conditioning drawing of a project should be closely checked against structural drawings and the field conditions to assure:

_____ 1. Clearance between heating, ventilating, and air-conditioning items and structural:
   a. Floor slab, mezzanines, platforms, and canopies
   b. Girders, beams, and tie rods
   c. Monorail steel
   d. Framing for stairs, elevators, and escalators
   e. Columns and column pads
   f. Footings and piers
   g. Struts, braces, and clips
   h. Special equipment supports and foundations

_____ 2. Provision of:
   a. Extra floor framing for heating, ventilating, and air-conditioning piping and duct penetrations
   b. Extra lintels for heating, ventilating, and air-conditioning wall and partition penetrations
   c. Upset beams at duct shafts (when required)
   d. Fuel-tank anchoring slab and strap details and proper depth of adjacent footing
   e. Special reinforcement of building structure to accommodate heavy heating, ventilating, and air-conditioning equipment
   f. Special support details for cooling towers, compressors, pumps, fans, and suspended equipment
   g. Special subway-type grating catwalks for boilers and platforms for auxiliary equipment (when required)
   h. Thrust-block details for underground high-pressure piping
   i. All-inclusive typical details of sleeve locations of heating risers, especially along exterior walls (for prevention of obtrusive installations of piping in occupied areas)
   j. Special note to general contractor to the effect that all structural items involving heating, ventilating, and air-conditioning work must first be verified and coordinated with the heating, ventilating, and air-conditioning contractor before work on openings, pads, and supports begins
   k. Proper reinforcement detail for supporting holes in concrete-plank roofs

6.5: *(Continued)*

NOTES:

Plumbing Each heating, ventilating, and air-conditioning drawing of a project should be closely checked against the plumbing drawings and the field conditions to assure:

_____ 1. Clearance between heating, ventilating, and air-conditioning items and plumbing:
   a. Plumbing mains and valves
   b. Plumbing risers and valves
   c. Cleanouts and shock absorbers
   d. Roof drains and leaders
   e. Floor drains and sump pits
   f. Drain-system vent pipes and outdoor-air intakes
   g. Plumbing fixtures, towel dispensers, and waste-receptacle units

_____ 2. Provision of gravity or mechanical ventilation for interior toilets, shower rooms, janitor closets, etc.

_____ 3. Availability of adequate water pressure at humidifier nozzle

_____ 4. Provision of makeup-water connections for:
   a. Boilers
   b. Expansion tanks
   c. Spray systems
   d. Electrostatic filter sprinklers
   e. Cooling towers
   f. Water-cooled fan bearings, compressor bearings, etc.

_____ 5. Provision of properly located drains and hose-bib connections for:
   a. Boiler rooms
   b. Equipment rooms
   c. Cooking towers and evaporative condensers
   d. Expansion tanks
   e. Electrostatic filters
   f. Drain pans
   g. Cooling-coil condensate
   h. Water-cooled condensers and aftercoolers
   i. Pumps

NOTES:

Electrical Each heating, ventilating, and air-conditioning drawing of a project should be closely checked against electrical drawings and the field conditions to assure:

_____ 1. Clearance between heating, ventilating, and air-conditioning items and electrical:
   a. Lighting fixtures
   b. Wall outlets and switches

c. Horizontal conduits and risers
d. Panels (lighting, power, telephone, and signal)
e. Pull boxes
f. Switchboards, load centers, battery racks, etc.

_____ 2. Agreement of locations of:
a. Electric motors and motor starters
b. Disconnect switches and emergency switches
c. Remote-control pushbutton and panel stations
d. Automatic electric valve and damper motors
e. Solenoid valves and relays (electric, electric-pneumatic and pneumatic-electric)
f. Electric thermostats and humidstats
g. Pressurestats and control switches (electric, electric-pneumatic and pneumatic-electric)

_____ 3. Provision of an adequate number of:
a. Marine lights for fan casings and plenums
b. Lighting fixtures for equipment rooms, control panels, boilers, etc.
c. Wall receptacles in equipment rooms
d. Convenience outlets and lighting fixtures for outdoor installations of cooling towers, etc.
e. Safety signs indicating separately energized electric interlocks
f. Mounting boards and control panels

_____ 4. Provision of adequate wall and floor space for mounting electrical control equipment

NOTES:

FORM 6.5

Key Use of Form:

Checklist to be used when reviewing a set of heating, ventilating, and air-conditioning drawings.

Who Prepares:

Person reviewing the drawings.

Who Uses:

Person performing review.

How to Complete:

Complete general information and check off items as drawings are being reviewed.

Alternative Forms:

Form can be revised to be more company-specific.

6.6: Environmental Control System Pretesting and Startup Checklist

ENVIRONMENTAL CONTROL SYSTEM PRETESTING AND STARTUP CHECKLIST

System Designation ______________________ Location ______________________

Completed By ________________________________ Date ________________

Prestart Inspection The steps indicated below should be carried out with the system inoperative.

1. **Check for system cleanliness** Before installing clean filters and spray elements, etc., check all ductwork for cleanliness, with particular attention to the following items:
   - _____ A. Eliminator sections
   - _____ B. Air intake (screens and louvers)
   - _____ C. Humidifiers
   - _____ D. Control sensors (thermostats, humidistats, etc.)
   - _____ E. Washer sumps
   - _____ F. Dampers
   - _____ G. Fan internals
   - _____ H. Supply and return diffusers, registers, and grilles
   - _____ I. Heating and cooling coils
   - _____ J. Floor gulleys, sinks, and related drainage systems
   - _____ K. Ductwork, internally and as far as physically possible
   - _____ L. Fan and other equipment chambers
   - _____ M. Mixing boxes or other terminals
   - _____ N. Cooling-coil drip pans
2. **Visual check of air-regulating devices and other components normally installed within ductwork** The following items should also be checked:
   - _____ A. Splitters, dampers, turning vanes, thermal insulation, and acoustic linings properly secured and fitted
   - _____ B. Damper clearances
   - _____ C. Free movement to fire dampers
   - _____ D. Damper seating
   - _____ E. All dampers secured in open position with the actuator disconnected if motorized
   - _____ F. Freedom of damper movement
   - _____ G. Location, fitting, and access to fusible-link assemblies
   - _____ H. Freedom from damage to acoustic linings, coil fins, and sensing elements
   - _____ I. Setting of terminal air-distribution devices in anticipated positions
   - _____ J. Position of damper blades with respect to quadrant indication
   - _____ K. Pinning to damper spindles
   - _____ L. Relative position of blades in multiple-leaf dampers
3. **Visual check for airtightness** Check:
   - _____ A. Covering of test holes
   - _____ B. Seating of filter and washer cells
   - _____ C. Water seals filled

_____ D. Ductwork joints, including flexible couplings
_____ E. Inspection covers fitted
_____ F. Equipment-room door seals around entire periphery

4. **Fans** Check:
_____ A. All components, fittings, bolts, etc., secure
_____ B. Antivibration dampers and fixings
_____ C. Leveling and alignment of motor and fan shaft and slide rails
_____ D. Bearing coolant
_____ E. In-line fans installed for correct airflow direction
_____ F. Bearing cleanliness
_____ G. Securing of pulleys
_____ H. Guards with finger access for speed measurement of motor and fan
_____ I. Belt tension, matchings, and distortion absence due to prolonged pretension
_____ J. Correct drive fitted
_____ K. External cleanliness
_____ L. Fresh and correct grade of lubricant
_____ M. Impeller free to rotate, in static balance, of correct handling, and secured

5. **Automatic filters** Check:
_____ A. Clearance, free movement, and alignment
_____ B. Leveling
_____ C. Drive tension
_____ D. Lubrication
_____ E. Correct oil level and clean oil bath in gearboxes and filter baths

6. **Control linkages (to dampers, etc.)** Check:
_____ A. Tightness of locking devices
_____ B. Free movement
_____ C. Bearing lubrication
_____ D. Correct alignment
_____ E. Fit of pins
_____ F. Freedom from excessive lost motion
_____ G. Stiffness of members
_____ H. Rigidity of mountings

7. **Electrical checks** Isolate all power supplies and check the following:
_____ A. Motor clean, all passageways clear, and bearings lubricated
_____ B. Reduced-voltage selection, e.g., transformer taps
_____ C. Starters, ammeter ranges, and circuit breakers relative to motor circuitry and as listed on motor nameplate
_____ D. Correct motor fitted
_____ E. Looped or flexible electrical connections fitted and secured on motionless side of equipment secured by vibration isolators
_____ F. Setting of timers
_____ G. Power and control wiring complete and correct, local isolation provided
_____ H. Overload settings
_____ I. Next connect to the power source and check:
(1) Declared voltage available on all supply phases

6.6: *(Continued)*

(2) In the case of large motors or complex circuitry starter operation sequence, safety interlocks, and timers can be checked with the motors unloaded

8. **Precipitators** Check:

_____ A. Correct alignment of plates before and/or after next screen fitted
_____ B. Safety interlocks
_____ C. Ionizing wires in position
_____ D. Warning notices fitted
_____ E. All cells connected
_____ F. Electric air heaters (check thermal-reset correct and readily accessible)

9. **Start-up Procedure** Once the duct system and plant have been checked as described above, the following steps, in order, should be carried out before balancing:

_____ A. Verify the correct rotation of impeller
_____ B. Start fan and run continuously
_____ C. Check motor running current
_____ D. Check fan and motor speeds and sheave sizes
_____ E. Check the total volume of air handled
_____ F. Check the total static pressure at the fan
_____ G. Set total volume with 10 to 15 percent more air than design values (since it will be found that during balancing the systems, the damper settings will reduce the volume handled)
_____ H. Next check the differential pressure on the filter manometer against the known volume handled by the fan
_____ I. The system is now ready to be balanced

NOTES:

FORM 6.6

Key Use of Form:

Used as a guide to pretest and startup various environmental control systems.

Who Prepares:

Person performing activities.

Who Uses:

Person performing activities.

How to Complete:

Provide the general information in the appropriate spaces and check off items as they are performed.

Alternative Forms:

The form can be made more company-specific by making the appropriate revisions to it.

6.7: Equipment Survey

EQUIPMENT SURVEY

Unit ______________________ Date ______________________

---

1. AIR HANDLERS—PACKAGE UNITS

Overall Appearance	☐ Good	☐ Improvement Needed	
Filters	☐ Clean	☐ Dirty	
Cleaned/Replaced	☐ Monthly	☐ Bimonthly	☐ Quarterly
	☐ Semiannual	☐ Annual	
Coils	☐ Clean	☐ Dirty Date Last Cleaned ______	
Cleaned	☐ Semiannual	☐ Annual	
Dampers	☐ Good	☐ Corroded	
	☐ Operational	☐ Nonoperational	
Bearings/Belts/Motors	☐ Good	☐ Repairs Required	

Remarks: ______________________

____

2. PUMPS

Overall Appearance	☐ Good	☐ Improvement Needed
Bearings/Packings/Motors	☐ Good	☐ Repairs Required

Remarks: ______________________

____

3. TEMPERATURE CONTROL SYSTEM

Overall Appearance	☐ Good	☐ Improvement Needed	
Compressor Bled and Oil Changed	☐ Weekly	☐ Monthly	☐ Automatic
	☐ Semiannual	☐ Annual	
Dryer Maintained	☐ Yes	☐ No	

Remarks: ______________________

____

4. BOILER

Overall Appearance	☐ Good	☐ Improvement Needed
Water Treatment Program	☐ Yes	☐ No Company ______
Annual Maintenance Performed	☐ Yes	☐ No

Remarks: ______________________

____

6.7: *(Continued)*

5. COOLING TOWER

Overall Appearance	☐ Good	☐ Improvement Needed
Water Treatment Program	☐ Yes	☐ No Company ________
Cleaned	☐ Bimonthly	☐ Quarterly
	☐ Semiannual	☐ Annual
Belts/Gear Box/Motor	☐ Maintained	☐ Repairs Required

Remarks: ____

____

6. UNIT HEATERS

Overall Appearance	☐ Good	☐ Improvement Needed
Operational	☐ Yes	☐ No
Annual Maintenance Performed	☐ Yes	☐ No

Remarks: ____

____

7. CHILLERS

Overall Appearance	☐ Good	☐ Improvement Needed	
Logs Maintained	☐ Yes	☐ No	
	☐ Every 10 Hrs Operation		☐ Less Often
Annual Maintenance Performed	☐ Yes	☐ No	
Oil Analyzed	☐ Yes	☐ No	

Remarks: ____

____

8. EXHAUST/RELIEF FANS

Overall Appearance	☐ Good	☐ Improvement Needed
Bearings/Belts/Motors	☐ Good	☐ Repairs Required

Remarks: ____

____

9. AUTOMOTIVE SHOP EQUIPMENT

Overall Appearance	☐ Good	☐ Improvement Needed
Battery Testers/Chargers	☐ Good	Date Calibrated ________
	☐ Repairs Required	
Brake Equipment	☐ Good	☐ Repairs Required
Front End Machine	☐ Good	Date Calibrated ________
	☐ Repairs Required	

6.7: *(Continued)*

Lifts/Racks	☐ Good	☐ Repairs Required	
Tire Changers	☐ Good	☐ Repairs Required	
Wheel Balancer	☐ Good	Date Calibrated ______________	
	☐ Repairs Required		
Air Compressor— Oil Changed	☐ Monthly	☐ Quarterly	
	☐ Semiannual	☐ Annual	
Bled Daily	☐ Yes	☐ No	
System Air Leaks	☐ Yes	☐ No	

Remarks: ____

____

10. ENERGY MANAGEMENT

Operating	☐ Yes	☐ No	
Type	☐ Boiler	☐ Chiller	☐ Fans
	☐ HVAC	☐ Lighting	☐ HVAC/Lighting
Energy Management Manual Being Followed	☐ Yes	☐ No	

Remarks: ____

____

11. MAJOR REPAIR—REPLACEMENT

Budgeted Current Year ____

____

To Be Budgeted Next Year ____

FORM 6.7

Key Use of Form:

Used in the performance of an inspection of the equipment covered by the form.

Who Prepares:

Maintenance inspector.

Who Uses:

Plant engineering personnel to evaluate existing maintenance activities and plan and schedule future tasks.

How to Complete:

As the inspection is performed, check the appropriate boxes.

Alternative Forms:

Changes can be made to make this form more company-specific. Also, other forms in this section can be used.

6.8: Annual Heating and Ventilating Systems Inspection Checklist

ANNUAL HEATING AND VENTILATING SYSTEMS INSPECTION CHECKLIST

Building ______________________ Date of Inspection ______________________

Maintenance Inspector's Name ____

EXTERIOR

The inspection of the exterior heating system consists of carefully noting the location and condition of the scheduled items.

Items to look for are:

1) Vandalism.
2) Badly rusted equipment and supporting devices requiring complete cleaning, priming and painting.
3) Masonry (chimneys) that is missing or masonry requiring tuck pointing of joints. Also masonry that is badly deteriorated by weathering or other action.
4) Damaged oil filling stations—including pipe and fittings.
5) Damaged vent piping and fittings.
6) Equipment and/or supporting devices so badly rusted or broken requiring replacement of parts.
7) Oil tank corrosion.
8) Bent and otherwise damaged louvers, bird screens, exterior dampers, ventilation equipment intake or exhaust ducts.
9) Inoperable exterior dampers.
10) Roof or wall type fan operation:
    (a) Fan running;
    (b) Damper operating.
11) Relief dampers or relief vents condition.
12) Roof flashing condition at:
    (a) Roof fan units;
    (b) Duct openings;
    (c) Pipe openings.
13) Chimney condition including supports.
14) Chimney weather cap conditions.

Exterior	Remarks	Cost Est.
1. Chimney and Gas Vents		
Masonry		
Metal		
Cap		
Roof flashing		
2. Roof Fan		
Fan operable		
Damper		

6.8: *(Continued)*

Exterior	Remarks	Cost Est.
Roof flashing		
Bird or insect screen		
General condition		
3. Roof Relief Openings		
Roof flashing		
Damper		
General condition		
4. Duct Intake or Exhaust		
General condition		
Roof flashing		
Bird or insect screen		
External damper		
Louvers		
5. Oil Tank		
General condition		
Sludge		
Fill piping		
Vent piping		
6. Gas		
Pipe condition		
Valves		
Vent pipe		
7. Wall Fan		
General condition		
Fan running		
Damper		
Flashing		
Bird or insect screen		
8. Unit Ventilator		
Exterior louver		

6.8: *(Continued)*

INTERIOR

The inspection of the interior heating system consists of carefully noting the location and condition of the scheduled items. Special note should be taken of noisy operation, leaks, excessive dirt accumulation, nonuniform temperatures, bearing failure.

Items to look for are:

1) Vandalism.
2) Inoperative equipment.
3) Damaged equipment.
4) Missing equipment.
5) Excessive noise.
6) Excessive vibration.
7) Excessive dirt, rust, or scale.

Where equipment has movable parts, they should be operated. Burners and stokers should be operated and all operating and safety controls checked for proper working condition. Pumps, fans, heating units, heating and ventilating units, etc., should all be operated by means of their normal control devices.

Controls should be checked by competent mechanics on a twice-a-year basis. Maintenance contracts with a control manufacturer are desirable. Electrical and gas apparatus should be continually checked for deterioration, leaks, etc.

Indicate the condition of the various items, location, or other identification, type of service, repair or replacement needed, estimated cost if possible.

Interior	Remarks	Cost Est.
1. Boiler		
General condition		
Smoke vent and/or breeching		
Draft control damper		
Insulation		
Relief valves		
Boiler tubes		
Burner or stoker		
Controls		
Boiler feeder		
Water glass		
Combustion air intake		
Blow-down		
Water treatment		
2. Pumps		

6.8: *(Continued)*

Interior	Remarks	Cost Est.
Boiler feed pump		
General condition		
Motor		
Valves		
Tank		
Insulation		
Controls		
Pressure valve		
Condensate pump		
General condition		
Motor		
Valves		
Receiver		
Controls		
Circulating water pump		
General condition		
Motor		
Valves		
Controls		
3. Expansion Tank		
General condition		
Site glass		
Water level		
4. Valves		
Leaking		
Packing		
Ease of operation		
5. Traps		
Float		
Seat		
Orifices		

6.8: *(Continued)*

Interior	Remarks	Cost Est.
6. Convectors and/or Baseboard Radiation		
General condition		
Coils, dirty		
Valves		
Trap		
7. Forced Flow Convectors or Fan-Coil Units		
General condition		
Coils		
Controls		
Valves		
Motor		
Belt		
Bearings		
Trap		
Fan		
8. Unit Ventilator (Steam, Water, or Gas)		
General condition		
Coils, dirty		
Filter, dirty		
Motor		
Belt		
Bearings		
Fan		
Trap		
Dampers		
Controls		
Valves		
Outside air intake		
Combustion air intake		

6.8: *(Continued)*

Interior	Remarks	Cost Est.
Burner		
Gas valves and regulator		
Pilot and heat exchanger		
Vent and flashing		
9. Unit Heater (Steam, Water, or Gas)		
General condition		
Coil, dirty		
Motor		
Fan		
Bearings		
Valves		
Trap		
Burner		
Controls		
Belts		
Draft inducer		
Gas valve and regulator		
Pilot		
Burner and heat exchanger		
Gas vent and flashing		
10. Furnace		
General condition		
Filter, dirty		
Motor		
Belt		
Bearing		
Dampers		
Burner		
Controls		
Fan		
Gas valve and regulators		

6.8: *(Continued)*

Interior	Remarks	Cost Est.
Gas vent and flashing		
Outside air intake		
Combustion air intake		
11. Ventilating Units		
General condition		
Filter, dirty		
Motor		
Belt		
Fan		
Bearings		
Trap		
Valves		
Dampers		
Controls		
Burner		
Coils, dirty		
Outside air intake		
Roof flashing		
Gas valve and regulators		
Pilot		
Burner and heat exchanger		
Gas vent and flashing		
Combustion air intake		
12. Control System		
General condition		
Compressor		
Air filter—dryer		
Thermostats		
Valves and valve operators		
Summer–winter changeover switch		
Tubing or wiring condition		

6.9: Annual Plumbing Inspection Checklist

ANNUAL PLUMBING INSPECTION CHECKLIST

Building ______________________ Date of Inspection ______________________

Maintenance Inspector's Name ______________________________________

Exterior	Remarks	Cost Est.
1. Storm Drainage		
Catch basin		
Broken or missing grate		
Filled with rubbish		
Drain pipe		
Clogged		
Broken		
Areaways		
Filled with rubbish		
Broken drain		
Ground hydrants		
Sill cocks		
Downspouts		
Water valve boxes		
Gas piping		
Grease interceptor		
2. Other		

Interior	Remarks	Cost Est.
1. Corridor		
Cleanouts		
Drinking fountains		
Electrical water coolers		
2. Toilet Room		
Water closet		
Water closet seats		
Water closet valves		

6.9: *(Continued)*

Interior	Remarks	Cost Est.
Water closet tanks		
Lavatory		
Faucet		
Drain		
Urinal		
Valve		
Tank		
Drain		
Soap		
Dispensers		
Tank		
Floor drain		
3. Kitchen		
Sink		
Faucet		
Drain		
Disposer		
Gas valves (check pilots if shut off)		
Range		
Dishwasher		
Booster heater		
Gas pilot		
Temperature (190°F)		
Flue venting		
Faucets		
Floor Drains		
4. Home Economics		
Gas valves and pilots		
Sink		
Faucet		
Drain		
Disposer		

6.9: *(Continued)*

Interior	Remarks	Cost Est.
Washer		
Faucet		
Drain		
5. Science		
Gas		
Air		
Sinks		
Faucets		
Drains		
6. Shop		
Gas		
Air		
Gasoline and oil interceptor		
Drains		
7. Elevator		
Drain		
Sump pump		
8. Equipment Rooms		
Drain from cooling units		
Valves		
Drains		
Sump pumps		
Sewage ejectors		
Water heaters		
Circulating pumps		
Thermometers		
Water meter		
Gas piping		
9. Roof		
Drains and downspouts		

6.10: Annual Fire Protection Equipment Inspection Checklist

ANNUAL FIRE PROTECTION EQUIPMENT INSPECTION CHECKLIST

Building ______________________ Date of Inspection ______________________

Maintenance Inspector's Name ____

Fire Protection Equipment	Remarks	Cost Est.
1. Sprinkler Systems		
Valves:		
Must be O.S. and Y. type.		
Must be open.		
Corrosion repairs.		
Flow alarms:		
Required for system with 5 or more sprinkler heads. UL approved.		
Must activate building fire alarm system.		
Dry pipe systems:		
Check for proper air pressure in system.		
Check low air pressure alarm system. Should actuate fire alarm system trouble signal.		
Pressure or storage tanks:		
Check for proper pressure.		
Check for proper water level.		
Check low pressure or low water alarm system. Should actuate fire alarm system trouble signal.		
Distribution piping and sprinkler heads:		
Check for leaks and corrosion.		
Check for adequate hangers.		

6.10: *(Continued)*

Fire Protection Equipment	Remarks	Cost Est.
2. Standpipe and Hose Equipment		
Hose and hose rack or cabinet:		
Is hose properly stored and readily accessible?		
Is hose attached to standpipe with nozzle open and ready for use?		
Is hose serviceable without leaks, decay, etc.?		
3. Fire Extinguishers		
Check hose, nozzle, gaskets and examine for deterioration.		
Check for injuries due to misuse.		
Test pump action type to see if pump functions.		
Test continuous-stream type by pumping up pressure and ejecting vaporizing liquid in both up-and-down and around-the-clock positions.		
Test stored-pressure and CO_2 by discharging small amount of liquid or foam to atmosphere.		

6.11: Detailed Air-Conditioning System Inspection Checklist

DETAILED BUILDING AIR-CONDITIONING SYSTEM INSPECTION CHECKLIST

Date of Inspection ______________________________

Maintenance Inspector's Name ____

System Designation and Location ____

Instructions: Check off items after performing inspection and/or maintenance tasks.

_____ Lubrication: inadequate lubricating instructions, excessive bearing temperature; inadequate lubrication of bearings and moving parts, low oil level, poor oil condition. Lubricate as required; add lubricant if grease dispenser or oil cup is less than half full; add oil to crankcase of refrigerant compressor if below correct level and change if dirty; clean clogged oil lines.

_____ Rust and Corrosion: damage from rust and corrosion. Remove rust, paint bare spots and corroded areas, where applicable.

_____ Motors, Drive Assemblies, and Fans: dust, dirt, grease, accumulations; worn, loose, missing, or damaged connections and connectors; bent blades; worn or loose belts; unbalance, misalignment, excessive noise and vibration, end play of shafts, ineffective sound isolators, poor condition of motor windings and brushes. Remove accumulations; tighten loose connections and parts; tighten loose belts or replace multiple belts in sets when one is worn; replace defective brushes; make other minor repairs and adjustments.

_____ Wiring and Electrical Controls: loose connections, charred, broken or wet insulation, short circuits, loose or weak contact springs, worn or pitted contacts, defective operation, wrong fuses, other deficiencies. Tighten loose connections and parts; replace cords having wet insulation or where broken in two or more places, or braid that is frayed more than six inches; replace or adjust contact springs; clean contacts; replace defective or improper fuses; make minor repairs.

_____ Temperature and Humidity Controls: improper setting, loose connections, worn, dirty, pitted or misalignment of contacts, defective operation noted in observing operation through complete cycle, inaccuracy of thermostats found by comparing with mercury thermometer (dry-bulb type), inaccuracy of humidistats found by comparing with sling psychrometer and psychrometric chart. Adjust settings; tighten connections; clean contacts and adjust alignment.

_____ Steam and Hot Water Heating Units: clogging, dirty heat transfer surfaces, leaking, loose connections and parts, bent fins, misalignment, water hammer, air-binding, nonuniform heat spread, open bypass valves, below normal temperature readings, defective valves, traps, and strainers. Clean heat-transfer surfaces; tighten loose connections and parts; replace leaking valve packing; close bypass valves; clean valves, traps, and strainers.

_____ Electrical Heating Units: burned, pitted, or dirty electrical contacts, short-circuited sections of elements, low voltage in electrical circuits, dirty reflective heat-transfer surfaces. Clean contacts and heat-transfer surfaces; tighten loose connections; make minor repairs and adjustments.

6.11: *(Continued)*

_____ Air Ducts, Dampers, Registers, Grills, Louvers, and Bird and Insect Screens: soot, dirt, dust, and other deposits, leaks, broken, loose, or missing connections and parts, excessive vibration, material defects, defective operation of movable parts, hinge-parts failure, improper seasonal or operating settings of dampers, inadequate air distribution in branch circuits. Remove deposits; tighten or replace defective connections and parts; caulk around flashings and make weathertight; adjust damper settings; make minor repairs and adjustments.

_____ Thermal Insulation and Vapor Barriers: wet, damaged, or missing, broken tie-wires, loose bands, torn canvas jackets. Repair or replace to restore insulating properties.

_____ Air Filters: dust, grease, other deposits, missing parts, improper fit. Replace dirty throwaway type filters, and those missing or with improper fit; wash permanent-type filters in soap suds or solvents, rinse with hot water; restore viscous coating in accordance with manufacturer's instructions.

_____ Guards, Casings, Hangers, Supports, Platforms, and Mounting Bolts: loose, broken, or missing parts and connections, deformations, improper level, ineffective sound isolators. Tighten loose connections and parts; adjust level; replace defective sound isolators; make minor repairs and replacements.

_____ Pump Units: dust, dirt, other deposits, leaks, noise, vibration, loose or missing connections or parts, defective operation. Clean; repair leaks; make minor repairs.

_____ Piping: leaks, corrosion, deformations, materials defects of fittings, copper tubing, steel piping. Repair leaks; make minor repairs.

_____ Water Sprays, Weirs, and Similar Devices: external scale, leakage, defective valves including float valve in sump, clogged nozzles or pipes; improper positioning of spray nozzles, baffles, or eliminators to control spray drift, to compensate for prevailing winds, or to provide cooling for entire coil; material defects. Remove external scale; wipe external surfaces with cloth dipped in solvent; tighten leaky connections; replace defective gaskets; adjust float valve; clean clogged openings with wire; position nozzles and adjust baffles or eliminators; make other minor repairs.

_____ Compressors: dirt, dust, leakage of oil, water or refrigerants; loose connections, loose or worn belts or parts, misalignment, excessive noise and vibration, incorrect suction and discharge pressure. Remove dirt, dust, and other accumulations; use Halide torch to check refrigerant leaks and watch for flame color change indicating leak; tighten loose connections, belts and parts; replace worn belts; correct misalignment; make other minor repairs.

_____ Shell- and Tube-Type Condensers: dust accumulations, leaks including connection to cooling tower. (Internal scale indicated by small difference in temperature of refrigerant and cooling water when taken by mercury thermometer at inlets and outlets.) Remove dust accumulations.

_____ Self-Contained Evaporative Condensers: leaks at pump or in piping, improper setting of float control device, improper overflow of solids contained in water, insufficient outdoor air flow, clogged nozzles, inadequate spread of water. Repair all leaks; adjust float controls; adjust outdoor air flow; clean clogged nozzles; adjust sprays.

6.11: *(Continued)*

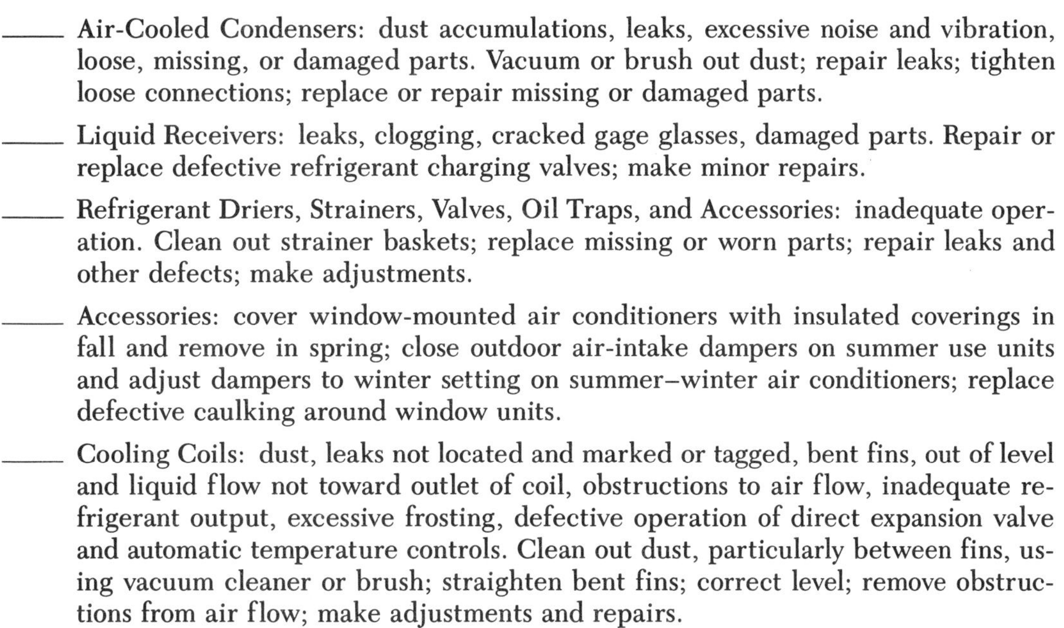

_____ Air-Cooled Condensers: dust accumulations, leaks, excessive noise and vibration, loose, missing, or damaged parts. Vacuum or brush out dust; repair leaks; tighten loose connections; replace or repair missing or damaged parts.

_____ Liquid Receivers: leaks, clogging, cracked gage glasses, damaged parts. Repair or replace defective refrigerant charging valves; make minor repairs.

_____ Refrigerant Driers, Strainers, Valves, Oil Traps, and Accessories: inadequate operation. Clean out strainer baskets; replace missing or worn parts; repair leaks and other defects; make adjustments.

_____ Accessories: cover window-mounted air conditioners with insulated coverings in fall and remove in spring; close outdoor air-intake dampers on summer use units and adjust dampers to winter setting on summer–winter air conditioners; replace defective caulking around window units.

_____ Cooling Coils: dust, leaks not located and marked or tagged, bent fins, out of level and liquid flow not toward outlet of coil, obstructions to air flow, inadequate refrigerant output, excessive frosting, defective operation of direct expansion valve and automatic temperature controls. Clean out dust, particularly between fins, using vacuum cleaner or brush; straighten bent fins; correct level; remove obstructions from air flow; make adjustments and repairs.

6.12: Piping System Inspection Checklist

PIPING SYSTEM INSPECTION CHECKLIST

Maintenance Inspector's Name ______________________________

Description and Location of Items Being Inspected ______________________________

____

Instructions: Perform the specified inspection and maintenance tasks. Insert the date when the activity is complete. Make any comments which are pertinent to future maintenance needs.

Item to Be Inspected	Date	Comments
Inspect all exposed piping for leakage, corrosion, loose connections, and damage. Tighten, repair, replace, and clean as required.		
Inspect all underground piping for leakage, ponding, erosion, or settlement.		
Check buried valves for bent stem, leakage, corrosion, and proper operation. Lubricate stem threads and packing as required.		
Check exposed valves for bent stem, leakage, corrosion, and proper operation. Lubricate stem threads and packing as required.		
Inspect meters for accuracy, leakage, corrosion, broken glass, moisture behind glass, settling, and proper operation. Adjust, repair, or replace as required.		
Inspect hydrants and hydrant shutoff valves (SOV) for missing caps, broken or missing chains, damaged threads, missing or damaged guards, and identification markings. Lubricate hydrant stem threads; replace or repair as required.		
Check hydrants for rust and corrosion. Remove rust and corrosion and apply paint where applicable.		

6.12: *(Continued)*

Item to Be Inspected	Date	Remarks
Check valve and meter pit manholes and roadway boxes for rust, corrosion, and damage. Remove rust and corrosion and apply paint where applicable. Repair or replace damaged parts as required.		
Check manhole frames, covers and ladder rungs for rust, corrosion, loose or missing rungs, and other damage or deficiencies.		
Check concrete and mortar joints in manholes.		
Check walkways, guardrails, stairs and ladders for rust, corrosion, broken or missing parts. Remove rust and corrosion and apply paint where necessary. Repair or replace parts as required.		
Check overflow pipes for water entrance, rust, and damage to screen. Remove rust and repair or replace screen as required.		

Provide any suggestions or recommendations for follow-up activities or any other comments related to the inspection and maintenance activities.

6.13: Water Heater (All Types) Inspection Checklist

WATER HEATER (ALL TYPES) INSPECTION CHECKLIST

Date of Inspection ____________________

Maintenance Inspector's Name ____________________

Water Heater Designation and Location ____________________

Instructions: Check off items after performing inspection and/or maintenance tasks.

_____ Lubrication: lubricate motor of oil burner and moving parts of mechanical devices.

_____ Rust and Corrosion: damage from rust or corrosion. Remove rust, paint bare spots and corroded areas, where applicable.

_____ Automatic Controls: improper operation through complete cycle; improper "on" and "off" operation. Check accuracy of thermostats by sling psychrometer or by wet-bulb or dry-bulb mercury thermometer; make adjustments required to attain normal performance.

_____ Safety and Flame Failure Devices: unsatisfactory as disclosed when tested through complete "on" and "off" cycle of operation.

_____ Thermal Insulation and Protective Coverings: open seams, breaks, missing sections, missing or loose fastenings. Replace or tighten fastenings; make minor repairs.

_____ Burner Assemblies: loose, damaged, or missing connections and parts, leakage, improper fuel-air mixtures, incorrect height and position of pilot light, improper baffle adjustment causing impingement, dirty heat-transfer surfaces, nonuniform flame spread, misalignment, clogged jets, orifices, and valves, dirty oil filters, defective oil wicks, oil rings, and pots. Test heating output at each step or multiple-step heating levels; tighten loose connections; repair or replace damaged or missing parts; adjust fuel-air mixture to produce blue flame; adjust pilot light; replace black or burnt-out pilot lamps; reset baffles; clean heat-transfer surfaces and ignition devices; correct misalignment; clean openings in jets, orifices, and valves; replace dirty oil filters, defective oil burner, other parts of assemblies.

_____ Combustion Chambers: deformations, breaks, cracks, wear, water and flue gas leakage, burnt-out grates, defective coal feed mechanisms, broken latches and hinges, door misalignment and poor fit, soot deposits, clinkers, ashes. Tighten loose connections; clean ash pits and grates; make minor repairs and replacements or adjustments.

_____ Electrical Heating Elements and Controls: loose connections, charred, frayed or broken insulation, short circuits, loose or weak contact springs, worn or pitted contacts, defective operation, wrong fuses, other deficiencies. Tighten loose connections; replace or adjust contact springs; clean contacts; replace defective fuses; make other minor repairs.

_____ Gauges: remove and calibrate gauges, instruments, and protective devices. Use standard thermometers and gauges if proper test facilities are not available.

6.13: *(Continued)*

_____ Water Compartments and Tanks: leaks, loose connections, chipped enamel finish, cracked cement linings, exposed bare metal or interior surfaces, defective manhole gaskets, dripping, corrosion, damage from freezing, defective operation of drain valves; deteriorated anodic rods or other devices provided for limiting corrosion. Tighten connections where feasible; make minor repairs; replace deteriorated anodic rods or segments.

_____ Water Relief and Steam Safety Valves: rust, corrosion, scale, mechanical defects, defective operation.

_____ Supports: unstable, material defects, loose, missing, or broken parts.

_____ Steam Coils and Instantaneous Water Heaters: improper steam pressure and water temperature, stream trap blowing, mechanical defects, clogged strainers, scaled heat-transfer surfaces, open bypass valves, leaking pipe connections.

_____ Hydrostatic Pressure Test: Gag or remove relief valves and cover the openings with blind flanges or pipe caps. Remove drain valves and other devices not designed for the test pressure. Do not perform hydrostatic tests if the surrounding atmosphere is at a freezing temperature. Task tank at 1.5 times maximum allowable operating pressure, holding the test pressure for not less than 2 hours. Drain water from tank after the test.

COMMENTS:

6.14: Gas Distribution System Inspection Checklist

GAS DISTRIBUTION SYSTEM INSPECTION CHECKLIST

Date of Inspection ______________________________

Maintenance Inspector's Name ______________________________

Gas System Designation and Location ______________________________

Instructions: Check off items after performing inspection and/or maintenance tasks.

_____ Exposed Piping: leakage, loose connections, rust, corrosion, other damage (use soap solution for leakage test).

_____ Underground Piping: signs of leakage such as brown grass strips across lawns, dead trees and shrubs when other plants are green in areas of buried piping. Determine condition when buried piping and valves are exposed for alteration and repair.

_____ Pressure-Regulating and Reducing Valves: leakage, loose connections, rust, corrosion, defective operation; possible damage from freezing resulting from water infiltration into valve pits; clogged vent to outside of building, defective screen.

_____ Gas Cutoff Valve on Face of Building: improper operation, difficult access, not clearly visible.

_____ Meters: loose connections, leakage, corrosion, rust, broken glasses, moisture behind glasses, defective gaskets, dirty or difficult to read, settlement.

_____ Pits: debris, water, cracks and leakage, rust and corrosion. Clean debris and water from pit; caulk all cracks; clean and paint all exposed parts, except moving parts.

COMMENTS:

6.15: Building Gas Heating System Inspection Checklist

BUILDING GAS HEATING SYSTEM INSPECTION CHECKLIST

Date of Inspection ______________________________

Maintenance Inspector's Name ______________________________

System Designation and Location ______________________________

Instructions: Check off items after performing inspection and/or maintenance tasks.

_____ Users' Comment: Ask occupants of building or system operator for comments on performance of heating system before starting inspection, note comments below.

_____ Lubrication: excessive bearing temperatures, inadequate lubrication of bearings and moving parts. Lubricate as required, including grease dispenser or oil cup when less than half full. Do not overlubricate.

_____ Rust and Corrosion: damage from rust and corrosion. Remove rust, paint bare spots and corroded areas, where applicable.

_____ Motor, Fans, Drive Assemblies, and Pumps: dust, dirt, other accumulations; leakage, wear, defective operation indicated from observation through operating cycle; loose, missing, or damaged connections and connectors, bent blades; worn or loose belts; unbalance, misalignment; excessive noise and vibration; end play of shaft; ineffective sound isolators; poor condition of motor windings and brush rigging. Remove accumulations; tighten loose connections and parts; tighten loose belts or replace multiple belts in sets when one is worn; replace defective brushes; make other minor repairs and adjustments.

_____ Wiring and Electrical Controls: loose connections; charred, frayed, broken, or wet insulation; short circuits; loose or weak contact springs; worn or pitted contacts; low voltage; defective operation; wrong fuses; other deficiencies. Tighten connections and parts; replace cords having wet insulation or where broken in two or more places, or braid that is frayed more than six inches; replace or adjust contact springs; clean contacts; replace defective or improper fuses; make minor repairs.

_____ Thermostats and Automatic Temperature Controls: improper operation through complete cycle; improper "on" and "off" operation. Check accuracy of thermostats by sling psychrometer or by wet-bulb or dry-bulb mercury thermometer; make minor repairs.

_____ Thermal Insulation and Protective Coverings: open seams, breaks, missing sections, missing or loose fastenings. Replace or tighten fastenings; make minor repairs.

_____ Piping System Identification: illegible, incorrect, improper, missing.

_____ Burner Assemblies: loose, damaged, or missing connections and parts; leakage; clogged jets, orifices, valves, and fuel supply lines; dirty wicks, oil rings, oil pots, oil filters, and heat-transfer surfaces; low-quality fuel, insufficient oil or gas pressures, low voltage, incorrect damper or thermostat settings, misalignment, nonuniform flame or heat spread, wrong fuel–air mixture, incorrect position of pilot light, improper baffle adjustment is causing impingement, defects in multiple-step heating device. Tighten loose connections; remove clogging; clean heat-transfer surfaces and ash pits; adjust fuel pressures; correct damper settings; adjust fuel–air

6.15: *(Continued)*

mixture to produce blue flame; position pilot light and baffles; make other minor repairs and adjustments.

_____ Combustion Chambers and Smokepipes: soot, dirt, coal deposits, other accumulations, abrasions, wear, deformations, misalignment; broken, loose, or missing parts of stokers, ash pits, grates, hinges, doors; lack of weathertightness of seams and joints, breaks in thermal insulation casings. Remove accumulations; stop any leakage of fuel gases; adjust flue dampers; make other minor repairs and adjustments.

_____ Registers, Grills, Dampers, Draft Diverters, Plenum Chambers, Supply and Return Ducts: soot, dust, and other deposits; clogging, deformations, broken, loose, or missing parts, loose seams and joints, breaks in vapor barriers, hinge parts failure, improper air distribution at branch ducts, improper seasonal damper or register settings. Remove deposits inside ducts; tighten loose connections; make minor repairs.

_____ Electrical Heating Units: burned, pitted, or dirty electrical contacts, short-circuited sections of elements, low voltage in electrical circuits, dirty reflective heat-transfer surfaces. Clean contacts and heat-transfer surfaces; tighten loose connections; make minor repairs and adjustments.

_____ Guards, Casings, Hangers, Supports, Platforms, and Mounting Bolts: loose, broken, or missing parts and connections, deformations, improper level, ineffective sound isolators. Tighten loose connections and parts; adjust level; replace defective sound isolators; make minor repairs and replacements.

_____ Steam and Hot Water Heating Equipment: dust, scale, other deposits, clogging, leaks, air-binding or water hammer, misalignment and improper slope of unit resulting in inadequate drainage and heating efficiency. Remove deposits; clean strainers; stop leaks at packing glands and repair other minor leaks; release entrapped air; make other minor repairs.

_____ Accessible Steam, Water, and Fuel Oil Piping and Valves (steam and condensate return, hot water heating, humidifier water supply, fuel oil systems): defective operation, leaks, clogging, casting blowholes, material defects, moisture, vibration, other faults that interfere with the proper operation. Repair or replace any defective valves; remove clogging; clean orifices; have seats of defective globe valves resurfaced and disks replaced, when valves or fittings are hazardous get permission to shut off; blow off moisture and entrapped air to relieve air-binding; repair minor leaks.

_____ Traps: leakage, defective operation. Disassemble and repair.

_____ Humidifier Assemblies: dust, leaking pans, solids in water, clogged piping, inoperative valves, danger of water overflow. Clean assemblies; removing clogging; adjust floats; add water to empty pans.

_____ Air Filters: dust, grease, other deposits, missing, improper fit. Replace dirty throwaway types and those missing or with improper fit; wash permanent types, restore viscous coating in accordance with manufacturer's instructions.

COMMENTS:

6.16: Paint Spray Booth Inspection Checklist

PAINT SPRAY BOOTH INSPECTION CHECKLIST

Maintenance Inspector's Name ______________________________

Description and Location of Items Being Inspected ______________________________

____

Instructions: Perform the specified inspection and maintenance tasks. Insert the date when the activity is complete. Make any comments which are pertinent to future maintenance needs.

Item to Be Inspected	Date	Comments
Safety—Comply with all current safety precautions.		
Check with operator for operating characteristics.		
Check operation of pumping unit and vent unit.		
Lubricate pump and fan shaft bearings. *Do not over lubricate.*		
Check hoses, pipes, and fittings for clogging and/or leakage.		
Inspect and adjust V-belts on vent units and replace if necessary.		
Inspect and replace starter contacts if necessary.		
Check coupling and shaft alignment.		
Check hold down bolts for tightness.		
Check pump and motor-shaft bearing for wear.		
Check filters and replace as required.		
Lubricate motor as required. *Do not over lubricate.*		
Inspect wiring and electrical controls for loose connections; charred, broken, or wet insulation; evidence of short circuiting and other deficiencies. Tighten, repair, replace as required.		

Provide any suggestions or recommendations for follow-up activities or any other comments related to the inspection and maintenance activities.

6.17: Septic Tank Inspection Checklist

SEPTIC TANK INSPECTION CHECKLIST

Date of Inspection ______________________________

Maintenance Inspector's Name ______________________________

Tank Designation and Location ______________________________

Instructions: Check off items after performing inspection and/or maintenance tasks.

GENERAL

_____ Manhole Frames and Covers: rust, corrosion, poor fit, missing, physical damage.

_____ Concrete and Masonry Surfaces: cracks, breaks, spalling, deteriorated mortar joints.

_____ Inlet and outlets: clogging, high concentration of suspended solids that may clog subsurface disposal fields.

_____ Flooding: wall surfaces above normal liquid levels show signs of frequent or occasional flooding. Determine source of infiltration on influent side.

SEPTIC TANKS

_____ Sediment: check depth. When sediment is two feet or less from effluent invert, septic tank should be pumped out; flush off concrete, and masonry surfaces; deposits of grease or oil scum indicate improper operation of interceptors.

DOSING TANKS

_____ Tank: If liquid level is less than three inches below level of liquid in septic tank, defective operation of siphon or clogged drainage field is indicated.

_____ Siphon: overflow clogged or blocked, unsatisfactory operation. Remove or break up and flush clean any material interfering with operation.

DISTRIBUTION BOXES

_____ Stopboards and Gates: improper functioning, undue leakage. Remove accumulations that might cause odors or interfere with proper seating.

_____ Water Levels: Water level above invert of outlet or slow drainage indicates poor functioning of drain fields.

TILE DISPOSAL FIELDS

_____ Ground Surface: ponding; indications of heavy trucking loads or other traffic that may break drains or force them out of alignment.

COMMENTS:

6.18: Earthquake Valve and Pit Inspection Checklist

EARTHQUAKE VALVE AND PIT INSPECTION CHECKLIST

Date of Inspection ______________________________

Maintenance Inspector's Name ____

Valve and Pit Designation and Location ____

Instructions: Check off items after performing inspection and/or maintenance tasks.

_____ Pit: (Sweep cover and surrounding area clean before removing cover.) debris, water, cracks, leakage, rust, corrosion.

_____ Valve: not absolutely level, leakage around all joints, gaskets, and cap plug on top.

_____ Prior to operational check and tripping of valve, examine appliances having pilot lights and record their location; open all windows in rooms where these appliances are located.

_____ Close remote shutoff valve, if so equipped, or tap lightly by hand only (never with metal object) until by listening close to valve, the sound indicates the valve pendulum has dropped and is not in closed position.

_____ Light gas burner at fixture nearest valve and test for leakage through valve. (If flame dies out after a few minutes, the valve is tight.)

_____ Remove cap plug at top of valve, and reset valve by lifting up on exposed stem.

_____ Replace plug, tighten, and test for leakage around plug.

_____ Make sure all pilot lights are relit.

DO NOT TAP OR JAR VALVE AND KEEP ALL WRENCHES AND METAL OBJECTS FROM MAKING CONTACT WITH VALVE.

COMMENTS:

6.19: Unit Heater Inspection Checklist

UNIT HEATER INSPECTION CHECKLIST

Maintenance Inspector's Name ______________________________

Description and Location of Items Being Inspected ______________________

____

Instructions: Perform the specified inspection and maintenance tasks. Insert the date when the activity is complete. Make any comments which are pertinent to future maintenance needs.

Item to Be Inspected	Date	Comments
Inspect wiring and electrical controls for loose connections; charred, broken, or wet insulation; short circuits; and tighten, repair, or replace as required.		
Lubricate electric motor as applicable.		
Lubricate fan shaft bearings as applicable.		
Check all bolts and brackets. Tighten as required.		
Inspect wall switch or thermostat and repair, replace, or adjust as required.		
Check water, steam, or gas lines for leaks.		
Check steam trap operation and repair as required.		
Clean burners as required.		
Check condition of and adjust tension on V-belt. Replace as required.		

Provide any suggestions or recommendations for follow-up activities or any other comments related to the inspection and maintenance activities.

6.20: Steam Trap Inspection Checklist

STEAM TRAP INSPECTION CHECKLIST

Maintenance Inspector's Name ______________________________

Description and Location of Items Being Inspected ______________________________

____

Instructions: Perform the specified inspection and maintenance tasks. Insert the date when the activity is complete. Make any comments which are pertinent to future maintenance needs.

Item to Be Inspected	Date	Comments
Check traps and bypass valves for leakage, damage, and proper operation.		
Check strainers for leakage, damage, proper operation, and clogged screens. Clean screens as required.		

Provide any suggestions or recommendations for follow-up activities or any other comments related to the inspection and maintenance activities.

6.21: Cooling Tower Inspection Checklist

COOLING TOWER INSPECTION CHECKLIST

Maintenance Inspector's Name ______________________________

Description and Location of Items Being Inspected ______________________________

____

Instructions: Perform the specified inspection and maintenance tasks. Insert the date when the activity is complete. Make any comments which are pertinent to future maintenance needs.

Item to Be Inspected	Date	Comments
Inspect wiring and electrical controls for loose connections; charring, broken, or wet insulation; evidence of short circuiting and other deficiencies. Tighten, repair, or replace as required.		
Check motor for excessive heat and vibration.		
Lubricate electric motor as applicable.		
Lubricate gearbox/drive shaft as applicable.		
Check all piping and connections. Tighten loose connections, replace gaskets and adjust float level as required.		
Check fan for bent blades, unbalance, excessive noise, and vibration.		
Inspect mounting brackets, bolts, etc., and tighten or replace as required.		
Check and inspect belts and adjust or replace as applicable.		
Check pulleys for alignment as applicable.		
Inspect for rust and corrosion. Remove rust and corrosion and apply paint where applicable.		
Check nozzles and diffusers. Adjust as required.		
Check reservoir for leaks and/or missing sealant. Repair as required.		

Provide any suggestions or recommendations for follow-up activities or any other comments related to the inspection and maintenance activities.

6.22: Small Turbine Inspection Checklist

SMALL TURBINE INSPECTION CHECKLIST

Date of Inspection ______________________________

Maintenance Inspector's Name ______________________________

Small Turbine Designation and Location ______________________________

Instructions: Check off items after performing inspection and/or maintenance tasks.

RUNNING INSPECTION

____ Vibration: excessive.

____ Bearings: excessive temperature.

____ Lubrication: dirty strainer, dirty or emulsified oil, improper level of oil or grease.

____ Oil Rings: improper operation.

____ Carbon Ring Seals: excessive steam leakage.

____ Constant Speed Machines: improper speed (check with tachometer).

____ Excess Pressure-Governed Turbines: improper pump discharge pressure.

____ Trip Valve: improper operation (overspeed unit and check speed with tachometer).

SHUTDOWN INSPECTION

____ Dismantle: remove turbine casing, bearing covers, governor housing, throttle valve, trip valve, and housings or covers that enclose moving parts.

____ Clearances and Moving Parts: dirty, no freedom of movement.

____ Rotor: remove deposits on blading.

____ Buckets and Blades: misalignment, corrosion, pitting, erosion. Clean, repair, or replace as required.

____ Carbon Rings or Other Types of Seals: excessive wear, breakage; journal in contact with carbon rings for wear and corrosion (particularly in units that have been idle for extended periods and those in which steam leakage has been noted around shaft).

____ Reassemble and perform running inspection as noted above.

COMMENTS:

6.23: Large Turbine Inspection Checklist

LARGE TURBINE INSPECTION CHECKLIST

Date of Inspection ______________________

Maintenance Inspector's Name ______________________

Large Turbine Designation and Location ______________________

Instructions: Check off items after performing inspection and/or maintenance tasks.

RUNNING INSPECTION

____ Vibration: excessive.

____ Lubrication: dirty or emulsified oil, improper level in sump, excessive temperature at bearing inlet and outlet, inadequate operation of emergency oil pump, inadequate operation of governor, throttle trip valve, and bleeder nonreturn tripping mechanism when lube oil pressure is low.

____ Steam Leakage: excessive steam leakage through carbon rings or labyrinth seals.

____ Pressure: improper stage pressures, improper gland seal pressure under varying loads.

____ Clearances: measure axial and radial clearances when built-in micrometer is installed.

____ Rankine (nonbleeding) Steam Rate: Check at rated load and half load, compare with manufacturer's standards.

____ Exhaust Casing Relief Valve: inadequate operation.

____ Varying Loads and Extraction Demands: inadequate operation of governor.

____ Reduced Load: inadequate operation of bleeder nonreturn valves.

____ No Load: inadequate operation of governor and trip valve when speed is raised to rpm of overspeed tripping mechanism (usually 10% above rated speed).

SHUTDOWN INSPECTION

____ Initial Clearances: Check clearances with integral micrometers, when installed, as soon as turbine has been taken off the line and rotor is still spinning.

____ Thrust-Bearing Clearance: Check by pushing rotor back and forth against stops.

____ Bearing Clearance: Remove bearing covers, check alignment with bridge gauge. (After unit has cooled, have lagging and upper half of casing removed.)

____ Blade Clearance: Check axial and radial clearances.

____ Balance Piston and Dummy Cylinder Clearance: Check axial clearance between balance piston and dummy cylinder.

____ Spindle: Check spindle for running true by rotating slowly in casing and using dial gauge on it at a number of points.

____ Blades: Remove spindle from casing, check for damaged edges, cracks on surfaces (by visual inspection, magnetic particle method, or black light method): erosion of last few rows (caused by moisture in steam at those stages); straighten edges and regauge blades as necessary; replace cracked or eroded blades.

6.23: *(Continued)*

_____ Diaphragms and Stationary Blade Rings: Remove diaphragms in impulse stage turbines and rings in reaction stage turbines from casings, check for damage and cracks, regauge and replace as needed, check sealing conditions and remove scale.

_____ Dummy Piston Sealing Rings and Runners: Check and replace as indicated.

_____ High-Pressure Interstage and Low-Pressure Packing: improper condition.

_____ Carbon Ring Packing: wear from wire drawing and breakage.

_____ Labyrinth Packing: signs of excessive rubbing on either rings or runners, evidence of abnormal clearance.

_____ Sealing Ring Seals: signs of leakage, excessive amounts of sealing steam required under operating conditions, corrosion and pitting of journals at sealing points.

_____ Cylinder Bore or Turbine Casing: evidence of warping, particularly the joining surfaces between upper and lower halves of casing; remove all high spots to assure perfect matching. (Extensive alterations may result in blade interference, particularly in reaction stages in which clearances are extremely close.)

_____ Speed Governor: Prior to dismantling, check clearance or backlash of driving gears; dismantle, check condition of all parts, such as pins, bushings, spindle bearings, weights, pivot valves, springs, and all links and levers that connect governor to pilot for servomechanism; check condition of servomotor piston, cylinder, and linkages to throttle bar; dismantle overspeed trip mechanism and check for freedom of motion and spring compression.

_____ Extraction Governor: Check condition of pressure-sensitive diaphragms, pilot linkages. Check condition of piston, rod, and cylinder of servomotor and all linkages between it and extraction control valve.

_____ Throttle Trip Valve: Prior to dismantling, determine time required for valve to become fully seated from instant of tripping hand lever; dismantle valve, check condition, take measurements of valve seat, guide, and bushings to determine amount of wear, and trueness of valve stem; valves showing signs of leaking valve seat should be ground in with valve in place.

COMMENTS:

6.24: Bakery Equipment Inspection Checklist

BAKERY EQUIPMENT INSPECTION CHECKLIST

Maintenance Inspector's Name ______________________________

Designation and Location of Item Inspected ______________________________

____

Instructions: Perform the specified inspection and maintenance tasks. Insert the date when the activity is complete. Make any comments which are pertinent to future maintenance needs.

Item to Be Inspected	Date	Comments
Ask operator or supervisor for comments on equipment performance.		
Inspect wiring and electrical controls for loose connections; charred, broken, or wet insulation; evidence of short circuiting and other deficiencies. Tighten, repair, or replace as required.		
Inspect motors, drive assemblies, and fans for flour, dirt, and grease; worn, broken, or missing parts and other damage. Clean, repair, or replace as required.		
Check effectiveness of sound insulators.		
Check motor for excessive heating and vibration.		
Lubricate motor as required. *Do not over lubricate.*		
Check operating controls for proper operation through complete operating cycle.		
Check meters, gauges, recording and indicating instruments, for accuracy and damage. Adjust, repair, or replace as required.		
Check fire protection devices for proper operation.		
Check thermal insulation and protective coverings for adequacy. Repair or replace as required.		
Check conveyor for proper speed, alignment, worn or damaged parts. Adjust, repair, or replace as required.		

Provide any suggestions or recommendations for follow-up activities or any other comments related to the inspection and maintenance activities.

6.25: Portable Water Pumping Plant Inspection Checklist

PORTABLE WATER PUMPING PLANT INSPECTION CHECKLIST

Maintenance Inspector's Name ______________________________

Designation and Location of Item Inspected ______________________________

____

Instructions: Perform the specified inspection and maintenance tasks. Insert the date when the activity is complete. Make any comments which are pertinent to future maintenance needs.

Item to Be Inspected	Date	Comments
Inspect valves for leaks and defects. Replace defective parts.		
Check pump for proper operation, leaks, and damage. Repair or replace as required. Lubricate as required.		
Check pump coupling for alignment and damage. Adjust, repair, or replace as required.		
Lubricate motor as required. *Do not over lubricate.*		
Check motor for excessive heat or vibration.		
Check supports for rust and corrosion. Remove rust and corrosion and apply paint.		
Inspect wiring and electrical controls for loose connections; charring, broken, or wet insulation; evidence of short circuiting and other deficiences. Tighten, repair, or replace as required.		

Provide any suggestions or recommendations for follow-up activities or any other comments related to the inspection and maintenance activities.

6.26: Fuel Distribution Facility Inspection Checklist

FUEL DISTRIBUTION FACILITY INSPECTION CHECKLIST

Maintenance Inspector's Name ______________________________

Designation and Location of Item Inspected ______________________________

____

Instructions: Perform the specified inspection and maintenance tasks. Insert the date when the activity is complete. Make any comments which are pertinent to future maintenance needs.

Item to Be Inspected	Date	Comments
Inspect all above-ground and underground piping for leaks, missing hangers or supports, defective gland nuts and bolts, and deterioration or damage to paint or protective coverings. Tighten, replace, or repair as required.		
Check valves for leaks, corrosion, damage or other deficiencies. Tighten, repair, clean, replace, and lubricate as required.		
Check meters and gauges for leaks, cracked or broken glass, defective gaskets, moisture, and accuracy of indicating and recording mechanism.		
Check accuracy of thermometers. Adjust or replace as required.		
Check strainers for leaks, obstructions, damage and wear. Clean, tighten, repair, or replace as required.		
Check shock arrester for leaks and proper operation.		
Check liquid in U-bend of liquid cushion arrester for proper level and add or remove liquid as required.		
Check vents for damaged or clogged screens. Repair, clean, or replace as required.		
Check all electrical ground connections for loose connections, electrical continuity, rust, and corrosion. Tighten, clean, repair, or replace as required.		

6.26: *(Continued)*

Item to Be Inspected	Date	Comments
Check grading at pits and tunnels.		
Check ladders for damage, missing or broken rungs, rust and corrosion. Repair, replace, or clean as required.		
Check all signs and markings for accuracy and legibility.		

Provide any suggestions or recommendations for follow-up activities or any other comments related to the inspection and maintenance activities.

6.27: Fuel Receiving Facility Inspection Checklist

FUEL RECEIVING FACILITY INSPECTION CHECKLIST

Maintenance Inspector's Name ______________________________

Designation and Location of Item Inspected ______________________________

____

Instructions: Perform the specified inspection and maintenance tasks. Insert the date when the activity is complete. Make any comments which are pertinent to future maintenance needs.

Item to Be Inspected	Date	Comments
Check platforms and islands for loose, missing, worn, or rotted planks. Tighten, repair, or replace as required.		
Check all connections for rust and corrosion and missing, broken, or damaged parts. Clean, repair, or replace as required.		
Check framing, supports, guardrails, and stairs for deterioration and damage. Repair or replace as required.		
Check hose racks and reels for rust and corrosion of metal components and for rotting and insect infestation of wood components. Clean, repair, or replace as required.		
Check signs and markings for accuracy and legibility.		
Check all grounding connections. Repair or replace as required.		
Inspect all painted surfaces for cracking, scaling, peeling, wrinkling and alligatoring, and loss of paint.		

Provide any suggestions or recommendations for follow-up activities or any other comments related to the inspection and maintenance activities.

6.28: Fuel Storage Facility Inspection Checklist

FUEL STORAGE FACILITY INSPECTION CHECKLIST

Maintenance Inspector's Name ______________________________

Designation and Location of Item Inspected ______________________________

____

Instructions: Perform the specified inspection and maintenance tasks. Insert the date when the activity is complete. Make any comments which are pertinent to future maintenance needs.

Item to Be Inspected	Date	Comments
Inspect foundations for settlement, cracking, and heaving.		
Inspect exterior concrete surfaces for cracks, exposed reinforcing, leaks, and spalling.		
Inspect exterior steel surfaces for rust, corrosion, and deteriorated paint.		
Inspect roof surfaces.		
Inspect floating and expansion-type roofs, seals, supports, and support guides for rust, corrosion, sealing, paint, and damage.		
Inspect structural supports and connections for rust, corrosion, rot, broken, cracked, distorted, loose, missing, and deteriorated paint.		
Inspect tank linings.		
Inspect tank interior.		
Inspect frames and covers on manholes and hatches for rust, corrosion, cracks, breaks, missing or damaged bolts, defective hinges and gaskets.		
Inspect vents for rust, corrosion, and dirty screens.		
Check pressure and vacuum relief valves for operation, leakage, and adjustment.		
Check manometers and thermometers for accuracy, damage, and level of fluid.		

6.28: *(Continued)*

Item to Be Inspected	Date	Comments
Check float gauges for wear, binding, and accuracy.		
Check cables, sheaves, and winch of swing lines for wear; damage; seals; operation.		
Inspect stairs, ladders, platforms, and walkways for rust, corrosion, rot; broken, cracked, loose, missing, members or connections; deteriorated paint.		

Provide any suggestions or recommendations for follow-up activities or any other comments related to the inspection and maintenance activities.

6.29: Walk-in Freezer Inspection Checklist

WALK-IN FREEZER INSPECTION CHECKLIST

Maintenance Inspector's Name ______________________________

Designation and Location of Item Inspected ______________________________

____

Instructions: Perform the specified inspection and maintenance tasks. Insert the date when the activity is complete. Make any comments which are pertinent to future maintenance needs.

Item to Be Inspected	Date	Comments
Replace dirty throwaway-type filter; wash permanent filters and restore viscous coating per manufacturer's instructions.		
Inspect wiring and electrical controls for loose connections; charring, broken, or wet insulation; evidence of short-circuiting and other deficiencies. Tighten, repair, or replace as required.		
Report or correct unsanitary conditions.		
Lubricate electric motor as applicable.		
Check motor for excessive heat and vibration.		
Inspect for rust and corrosion. Remove rust and corrosion, and apply paint where applicable.		
Check valves and pipes for leaks and loose connections. Repair, replace, or tighten as required.		
Check fans for freedom of rotation, bent blades, loose or missing parts, and repair, replace, or tighten as required.		
Check for proper guards, hangers, and supports. Tighten, repair, or replace as required.		
Clean coils as required.		

Provide any suggestions or recommendations for follow-up activities or any other comments related to the inspection and maintenance activities.

6.30: Sewage Collection and Disposal Systems Inspection Checklist

SEWAGE COLLECTION AND DISPOSAL SYSTEMS INSPECTION CHECKLIST

Maintenance Inspector's Name ______________________________

Designation and Location of Item Inspected ______________________________

____

Instructions: Perform the specified inspection and maintenance tasks. Insert the date when the activity is complete. Make any comments which are pertinent to future maintenance needs.

Item to Be Inspected	Date	Comments
GENERAL Inspect grease traps, oil interceptors, and similar equipment for accumulations of scum and grit.		
Inspect frame, cover, and ladder rungs for rust, corrosion, fit of cover, and damage.		
Inspect concrete and masonry for cracks, breaks, spalling, deteriorated mortar joints.		
Inspect piping for corrosion, open joints, cracked or crushed sections, obstructions.		
Inspect inverted siphons and depressed sewers for clogging, sluggish flow, accumulations of grit and debris.		
Check for leakage, rust, corrosion, deteriorated coatings. Remove rust, spot paint as required.		
Inspect supports and anchors.		
Check bar screen and raker for corrosion and damage.		
Inspect cutters.		
Check lubrication. *Do not over lubricate.*		
CHLORINATOR Inspect housing for adequate ventilation.		
Inspect assembly for rust, corrosion, or leaks. Remove rust or corrosion and spot paint as required.		

6.30: *(Continued)*

Item to Be Inspected	Date	Comments
VEGETATION AND ADJACENT GROUNDS Inspect grass and ground cover plants.		
Inspect ground surfaces for indications of seepage from sewers.		
TIDE GATE Check operation.		
Inspect for blockage.		
SPECIAL INSPECTION DURING OR AFTER PROLONGED RAIN OR SEVERE STORM Inspect manholes for infiltration, proper grading.		
Inspect underground piping.		
Inspect above-ground piping.		
Inspect vegetation on slopes for stabilization.		

Provide any suggestions or recommendations for follow-up activities or any other comments related to the inspection and maintenance activities.

6.31 Refrigeration Equipment Inspection Checklist

REFRIGERATION EQUIPMENT INSPECTION CHECKLIST

Maintenance Inspector's Name ______________________________

Designation and Location of Item Inspected ______________________________

____

Instructions: Perform the specified inspection and maintenance tasks. Insert the date when the activity is complete. Make any comments which are pertinent to future maintenance needs.

Item to Be Inspected	Date	Comments
Inspect wiring and electrical controls for loose connections; charred, broken, or wet insulation. Tighten, repair, or replace as required.		
Lubricate motor as required.		
Check motor for excessive heat and vibration.		
Check condition of belt and adjust or replace as necessary.		
Vacuum/clean dust from cooling coils as required.		
Check hermetic unit for housing leaks, evidence of overheating, excessive noise, and vibration.		
Inspect door gaskets and replace as required.		
Inspect for rust and corrosion. Remove rust and corrosion and apply paint where applicable.		

Provide any suggestions or recommendations for follow-up activities or any other comments related to the inspection and maintenance activities.

6.32: Safety Shower Inspection Checklist

SAFETY SHOWER INSPECTION CHECKLIST

Maintenance Inspector's Name ______________________________

Designation and Location of Item Inspected ______________________________

____

Instructions: Perform the specified inspection and maintenance tasks. Insert the date when the activity is complete. Make any comments which are pertinent to future maintenance needs.

Item to Be Inspected	Date	Comments
Check water flow.		
Check control valve for ease of operation.		
Check spray pattern from spray nozzle. Make appropriate adjustments as required to obtain proper spray pattern.		
Check drain. Clean as required.		
Check all pull chains, handles, bars, etc., for wear, alignment, and proper positioning. Repair or replace as required.		
Tag shower with date inspected and by whom.		

Provide any suggestions or recommendations for follow-up activities or any other comments related to the inspection and maintenance activities.

6.33: Swimming Pool Equipment Inspection Checklist

SWIMMING POOL EQUIPMENT INSPECTION CHECKLIST

Maintenance Inspector's Name ____________________

Designation and Location of Item Inspected ____________________

____

Instructions: Perform the specified inspection and maintenance tasks. Insert the date when the activity is complete. Make any comments which are pertinent to future maintenance needs.

Item to Be Inspected	Date	Comments
Check pump room for equipment leaks.		
Start pumps and check filters. Blow down if required.		
Check for unusual noise or vibration.		
Check pump and motor alignment; condition of coupling. Repair or replace as required.		
Lubricate motor as required. *Do not over lubricate.*		
Inspect for rust and corrosion. Remove rust and corrosion and apply paint where required.		
Inspect wiring and electrical controls for loose connections; charred, broken, or wet insulation; evidence of short-circuiting and other deficiencies. Tighten, repair, and replace as required.		

Provide any suggestions or recommendations for follow-up activities or any other comments related to the inspection and maintenance activities.

6.34: Swimming Pool Inspection Checklist

SWIMMING POOL INSPECTION CHECKLIST

Maintenance Inspector's Name ______________________________

Designation and Location of Item Inspected ______________________________

____

Instructions: Perform the specified inspection tasks. Insert the date when the activity is complete. Make any comments which are pertinent to future maintenance needs.

Item to Be Inspected	Date	Comments
Check concrete for cracks, breaks, spalling; exposed reinforcing; settlement.		
Check tile for chipped, cracked, loose, and missing pieces; mortor joints.		
Check expansion joints for leakage and damage.		
Check wall and floor finishes for roughness and dirt.		
Check depth markers and lane strips for legibility.		
Check springboards for cracks, breaks, splintering, and other damage; loose or missing fastenings; absence of nonslip coverings.		
Check ladders for rust or corrosion of metal parts; loose, missing, broken, rot, or other damage to wooden parts; alignment of towers.		
Check main drains for sediment and rust.		
Check gutter drains for obstructions.		
Check walls for stains.		
Check fences, barricades, dividing walls, and footings for broken, loose, missing, or other damage.		
Check painted surfaces for blistering, checking, cracking, scaling, wrinkling, flaking, peeling, rust, corrosion, absence of paint.		

Provide any suggestions or recommendations for follow-up activities or any other comments related to the inspection activities.

6.35: Chlorinator and Hypochlorinator Inspection Checklist

CHLORINATOR AND HYPOCHLORINATOR INSPECTION CHECKLIST

Maintenance Inspector's Name ____________________

Designation and Location of Item Inspected ____________________

____

Instructions: Perform the specified inspection and maintenance tasks. Insert the date when the activity is complete. Make any comments which are pertinent to future maintenance needs.

Item to Be Inspected	Date	Comments
Check to ensure that gas mask and ammonia water (ammonium hydroxide) bottle are in place.		
Check for leakage or corrosion on cylinders, piping, valves, and connections. Tighten or repair as required.		
Check for proper operation of scales, cylinder pressure gauge, or flow meter. Adjust or repair as required.		
Check all valves and metering devices for proper operation. Adjust, repair, or replace as required.		
Check pump for proper operation. Lubricate as required.		
Check strainer and clean as required.		
Inspect wiring and electrical controls for loose connections; charred, broken, or wet insulation; evidence of short-circuiting, and other deficiencies. Tighten, repair, or replace as required.		

Provide any suggestions or recommendations for follow-up activities or any other comments related to the inspection and maintenance activities.

6.36: Chemical Feed Equipment for Water Supply Inspection Checklist

CHEMICAL FEED EQUIPMENT FOR WATER SUPPLY INSPECTION CHECKLIST

Maintenance Inspector's Name ______________________________

Designation and Location of Item Inspected ______________________________

____

Instructions: Perform the specified inspection and maintenance tasks. Insert the date when the activity is complete. Make any comments which are pertinent to future maintenance needs.

Item to Be Inspected	Date	Comments
Check for leakage, loose bolts, and damage. Tighten, repair, or replace as required.		
Lubricate pump as required.		
Check pots or tanks for adequate supply of chemical, proper solution level, loose covers on dry-feed type, vibration or noise when operating, sediment in tank. Blow off sediment.		
Check piping for encrustation, potable water lines below rim of tank, air gap to prevent siphonage. Clean.		
Check for clogged or partially obstructed strainers. Clean as required.		
Check for proper operation of valves.		
Check sight glasses and rate-of-flow indicators for dirty glass and proper chemical feeding rate. Clean and adjust as required.		
Check water meters for proper operation.		
Check feed control units for proper operation.		
Check pumps for vibration, adequate belt tension, overheating of motor.		
Check motor for overheating and vibration.		

6.36: *(Continued)*

Item to Be Inspected	Date	Comments
Lubricate motor as required. *Do not over lubricate.*		
Inspect wiring and electrical controls for loose connections; charring, broken, or wet insulation; evidence of short-circuiting and other deficiencies. Tighten, repair, or replace as required.		

Provide any suggestions or recommendations for follow-up activities or any other comments related to the inspection and maintenance activities.

6.37: Eyewash Fountain Inspection Checklist

EYEWASH FOUNTAIN INSPECTION CHECKLIST

Maintenance Inspector's Name ______________________________

Designation and Location of Item Inspected ______________________________

____

Instructions: Perform the specified inspection and maintenance tasks. Insert the date when the activity is complete. Make any comments which are pertinent to future maintenance needs.

Item to Be Inspected	Date	Comments
Check control valve for proper operation.		
Check eyewash spray nozzles for proper spray pattern and pressure.		
Check drain for stoppage or leakage.		
Tag eyewash fountain with date inspected and by whom.		

Provide any suggestions or recommendations for follow-up activities or any other comments related to the inspection and maintenance activities.

6.38: Fresh Water Storage Inspection Checklist

FRESH WATER STORAGE INSPECTION CHECKLIST

Maintenance Inspector's Name ______________________________

Designation and Location of Item Inspected ______________________________

____

Instructions: Perform the specified inspection tasks. Insert the date when the activity is complete. Make any comments which are pertinent to future maintenance needs.

Item to Be Inspected	Date	Comments
Check foundations for settlement, damage, rot, and insect infestation.		
Check steel tanks for rust, corrosion, leakage, scale, damaged protective coating, and damage.		
Check concrete tanks for damage, cracks, spalling, leakage, and damaged protective coating.		
Check wood tanks for leakage, rot, insect infestation, and damage.		
Check towers for rust, corrosion, damage, rot, and insect infestation as applicable.		
Check expansion joints for loose or missing sealant.		
Check earth embankments for erosion, ponding of water, and leakage.		
Check valves, piping, fittings, and sleeves for rust, corrosion, leakage, and damage.		

Provide any suggestions or recommendations for follow-up activities or any other comments related to the inspection activities.

6.39: Ventilating Fan and Exhaust System Inspection Checklist

VENTILATING FAN AND EXHAUST SYSTEM INSPECTION CHECKLIST

Maintenance Inspector's Name ______________________________

Designation and Location of Item Inspected ______________________________

____

Instructions: Perform the specified inspection and maintenance tasks. Insert the date when the activity is complete. Make any comments which are pertinent to future maintenance needs.

Item to Be Inspected	Date	Comments
Lubricate fan shaft bearings and/or electric motors as applicable. *Do not over lubricate.*		
Check condition of belt and adjust or replace as necessary.		
Inspect wiring and electrical controls for loose connections; charred, broken, or wet insulation; short circuits; and tighten, repair, or replace as required.		
Check motor for excessive heat and vibration.		
Inspect for rust and corrosion. Remove rust and corrosion and apply paint where applicable.		
Clean screen and vent on roof as applicable.		
Check ducts, collectors, smokepipes, and hoods for clogging, soot, dirt, and grease. Perform minor cleaning operations as necessary.		
Inspect guards, supports, covers, etc., and tighten loose connections and posts.		

Provide any suggestions or recommendations for follow-up activities or any other comments related to the inspection and maintenance activities.

6.40: Air-Handling Unit Inspection Checklist

AIR-HANDLING UNIT INSPECTION CHECKLIST

Maintenance Inspector's Name ______________________________

Designation and Location of Item Inspected ______________________________

____

Instructions: Perform the specified inspection and maintenance tasks. Insert the date when the activity is complete. Make any comments which are pertinent to future maintenance needs.

Item to Be Inspected	Date	Comments
Inspect and change air filters as required.		
Lubricate fan shaft bearings and electric motors as required. *Do not over lubricate.*		
Check condition of V-belt(s), adjust tension or replace as required.		
Check alignment of V-belt pulleys; tighten set screws.		
Inspect wiring and electrical controls for loose connections; charred, broken, or wet insulation; evidence of short-circuiting; wrong size fuses; other electrical deficiencies; and tighten, correct, repair, or replace as required.		
Check motor for excessive heat and vibration.		
Check dampers for proper operation and lubricate as required.		
Check insulation and vapor barriers. Repair or replace as necessary to restore insulating and vapor barrier properties.		
Check and ensure that the evaporator coil drain is open and unit drains properly.		
Check and ensure that the required guards, supports, and mounting bolts are properly installed and secure.		

6.40: *(Continued)*

Item to Be Inspected	Date	Comments
Inspect for rust and corrosion. Remove rust and corrosion and apply paint where applicable.		
Inspect heating/cooling coils and clean as necessary to ensure maximum air flow.		

Provide any suggestions or recommendations for follow-up activities or any other comments related to the inspection and maintenance activities.

6.41: Liquid Filter Inspection Checklist

LIQUID FILTER INSPECTION CHECKLIST

Maintenance Inspector's Name ____________________

Designation and Location of Item Inspected ____________________

Instructions: Perform the specified maintenance tasks. Insert the date when the activity is complete. Make any comments which are pertinent to future maintenance needs.

Item to Be Inspected	Date	Comments
Clean inside of filter shell.		
Replace cartridges as required.		
Replace tubes and springs if necessary.		

Provide any suggestions or recommendations for follow-up activities or any other comments related to the maintenance activities.

6.42: Laundry Equipment Inspection Checklist

LAUNDRY EQUIPMENT INSPECTION CHECKLIST

Maintenance Inspector's Name ______________________________

Designation and Location of Item Inspected ______________________________

____

Instructions: Perform the specified inspection and maintenance tasks. Insert the date when the activity is complete. Make any comments which are pertinent to future maintenance needs.

Item to Be Inspected	Date	Comments
Inspect wiring and electrical controls for loose connections; charred, broken, or wet insulation; short circuits; wrong size fuses; other deficiencies; and tighten, correct, repair, or replace as required.		
Check condition of belts and/or chains and adjust or replace as required.		
Check alignment of pulleys and adjust as required.		
Lubricate bearings as applicable. *Do not over lubricate.*		
Check motor(s) for excessive heat and vibration.		
Check and ensure proper installation of guards, supports, and mounting bolts.		
Check for leaking pipe connections, valves, steam traps, and other mechanical defects; tighten, repair, or replace as required.		
Check gaskets/seals; repair or replace as required.		
Inspect for rust and corrosion. Remove rust and corrosion and apply paint where applicable.		
Check and clean lint filters and drains.		

Provide any suggestions or recommendations for follow-up activities or any other comments related to the inspection and maintenance activities.

6.43: Incinerator Inspection Checklist

INCINERATOR INSPECTION CHECKLIST

Maintenance Inspector's Name ______________________________

Designation and Location of Item Inspected ______________________________

____

Instructions: Perform the specified inspection and maintenance tasks. Insert the date when the activity is complete. Make any comments which are pertinent to future maintenance needs.

Item to Be Inspected	Date	Comments
Inspect wiring and electrical controls for loose connections; charring, broken, or wet insulation; evidence of short circuits and other deficiencies; tighten, repair, or replace as required.		
Lubricate electric motor(s) as applicable.		
Check cable and rigging of dampers as applicable.		
Inspect for rust and corrosion. Remove rust and corrosion and apply paint where applicable.		
Check operation of ash-handling and ash-removal facilities as applicable.		
Check motor(s) for excessive heat and vibration.		
Check gas or oil lines for leaks. Clean burners as required.		
Check air lines for leaks.		
Check combustion chamber.		
Check dampers and control mechanisms for proper operation.		

Provide any suggestions or recommendations for follow-up activities or any other comments related to the inspection and maintenance activities.

6.44: Ice Maker Inspection Checklist

ICE MAKER INSPECTION CHECKLIST

Maintenance Inspector's Name ____________________

Designation and Location of Item Inspected ____________________

____

Instructions: Perform the specified inspection and maintenance tasks. Insert the date when the activity is complete. Make any comments which are pertinent to future maintenance needs.

Item to Be Inspected	Date	Comments
Inspect door gaskets and replace as required.		
Inspect wiring and electrical controls for loose connections; charred, broken, or wet insulation; tighten, repair, or replace as required.		
Check control for proper setting. Adjust as necessary.		
Lubricate motor, pump, and drive mechanism as required.		
Check motor for excessive heat and vibration.		
Vacuum/clean dust from condenser coils as required.		
Clean ice coil and/or drum, storage bin, and drain.		
Adjust drum blades/ice collectors as required.		
Check hermetic unit for housing leaks, evidence of overheating, excessive noise, and vibration.		
Inspect for rust and corrosion. Remove rust and corrosion and apply paint where applicable.		

Provide any suggestions or recommendations for follow-up activities or any other comments related to the inspection and maintenance activities.

6.45: Heating and Ventilating Unit Inspection Checklist

HEATING AND VENTILATING UNIT INSPECTION CHECKLIST

Maintenance Inspector's Name ______________________________

Designation and Location of Item Inspected ______________________________

____

Instructions: Perform the specified inspection and maintenance tasks. Insert the date when the activity is complete. Make any comments which are pertinent to future maintenance needs.

Item to Be Inspected	Date	Comments
Change air filter as required.		
Check heater operation.		
Check and inspect belt; adjust tension as required.		
Lubricate shaft and motor bearings.		
Check wiring and electrical controls for loose connections; charred, frayed, or broken insulation; tighten, repair, or replace if necessary.		
Check for rust and corrosion. Remove rust and corrosion and apply paint where required.		

Provide any suggestions or recommendations for follow-up activities or any other comments related to the inspection and maintenance activities.

6.46: Heater and Console Control Inspection Checklist

HEATER AND CONSOLE CONTROL INSPECTION CHECKLIST

Maintenance Inspector's Name ____________________

Designation and Location of Item Inspected ____________________

____

Instructions: Perform the specified inspection and maintenance tasks. Insert the date when the activity is complete. Make any comments which are pertinent to future maintenance needs.

Item to Be Inspected	Date	Comments
Check airline lubricator and fill if necessary.		
Check lines to and from unit for steam, water, or oil leaks. Tighten or replace as required.		
Check dust filter for dust accumulation. Clean or change if necessary.		
Check alignment of pulleys and tighten set screws.		
Check condition of and adjust belt or replace as required.		
Lubricate electric motor bearings if applicable. *Do not over lubricate.*		
Check steam valves for proper operation. Repair or replace as required.		
Check steam trap. Repair or replace as required.		
Check motor bearings for noise or wear.		
Check base bolts for tightness.		
Inspect wiring and electrical controls for loose connections; charred, broken, or wet insulation; evidence of short-circuiting and other deficiencies. Tighten, repair or replace as required.		

Provide any suggestions or recommendations for follow-up activities or any other comments related to the inspection and maintenance activities.

6.47: Pump Vacuum Producer Inspection Checklist

PUMP VACUUM PRODUCER INSPECTION CHECKLIST

Maintenance Inspector's Name ______________________________

Designation and Location of Item Inspected ______________________________

____

Instructions: Perform the specified inspection and maintenance tasks. Insert the date when the activity is complete. Make any comments which are pertinent to future maintenance needs.

Item to Be Inspected	Date	Comments
Inspect wiring and electrical controls for loose connections; charred, broken, or wet insulation; short circuits; tighten, repair, or replace as required.		
Lubricate motor and/or pump as applicable.		
Check motor for excessive heat and vibration.		
Inspect for rust and corrosion. Remove rust and corrosion and apply paint where applicable.		
Check manufacturer's instructions for additional maintenance/service requirements.		
Check alignment and condition of belts. Adjust and replace as required.		

Provide any suggestions or recommendations for follow-up activities or any other comments related to the inspection and maintenance activities.

6.48: Dishwashing Equipment and Accessories Inspection Checklist

DISHWASHING EQUIPMENT AND ACCESSORIES INSPECTION CHECKLIST

Maintenance Inspector's Name ______________________________

Designation and Location of Item Inspected ______________________________

____

Instructions: Perform the specified inspection and maintenance tasks. Insert the date when the activity is complete. Make any comments which are pertinent to future maintenance needs.

Item to Be Inspected	Date	Comments
Inspect wiring and electrical controls for loose connections; charred, broken, frayed, or wet insulation; short circuits; defective operation and other deficiencies. Tighten, repair, and replace as required.		
Lubricate motor(s) as applicable. *Do not over lubricate.*		
Check conveyor(s) for proper speed, alignment, noise and vibration, loose chains or belts, defective sprockets, sheaves and rollers. Adjust chains, belts, and conveyor speed. Tighten connections and parts as required.		
Check motor(s) for excessive heat and vibration.		
Inspect pump for leaks, excessive heat and vibration, and defective operation. Repair or replace as required.		
Inspect garbage grinders for dull shredders or cutters, broken or loose parts, and excessive heat and vibration. Clean or replace strainer. Tighten, repair, or replace as required.		
Lubricate pump and garbage grinder as applicable.		
Inspect for rust and corrosion. Remove rust and corrosion and apply paint where applicable.		

6.48: *(Continued)*

Item to Be Inspected	Date	Comments
Check steam, water, drain or gas piping for loose connections, corrosion, and leaks. Tighten, repair, or replace as required.		
Inspect thermal insulation and protective coverings for open seams, missing sections, and loose fastenings. Tighten, repair, or replace as required.		
Check guards, casing, and covers for safe conditions.		
Inspect dishwashing compartment for leakage around door gasket, window or under machine, worn curtains and proper alignment of door. Repair, adjust, or replace as required.		
Check operating controls for proper operation and adjust as required.		
Check door seals/gaskets. Replace or repair as required.		

Provide any suggestions or recommendations for follow-up activities or any other comments related to the inspection and maintenance activities.

6.49: Dehumidification Unit Inspection Checklist

DEHUMIDIFICATION UNIT INSPECTION CHECKLIST

Maintenance Inspector's Name ______________________________

Designation and Location of Item Inspected ______________________________

____

Instructions: Perform the specified inspection and maintenance tasks. Insert the date when the activity is complete. Make any comments which are pertinent to future maintenance needs.

Item to Be Inspected	Date	Comments
Check desiccant level and condition. Add desiccant if necessary.		
Check banks for proper motor operation.		
Check and record amperage record of heater.		
Check times and damper linkage. Adjust as required.		
Check wiring and electrical controls for loose connections; charred, frayed or broken insulation. Tighten, repair, or replace as necessary.		
Lubricate motors as required.		
Check for rust and corrosion. Remove rust and corrosion and apply paint where required.		

Provide any suggestions or recommendations for follow-up activities or any other comments related to the inspection and maintenance activities.

6.50: Air Filter Inspection Checklist

AIR FILTER INSPECTION CHECKLIST

Maintenance Inspector's Name ____________________

Designation and Location of Item Inspected ____________________

____

Instructions: Perform the specified inspection and maintenance tasks. Insert the date when the activity is complete. Make any comments which are pertinent to future maintenance needs.

Item to Be Inspected	Date	Comments
Inspect filters for cleanliness.		
Replace disposable-type filters at frequency specified by manufacturer.		
Wash permanent-type filter at frequency specified by manufacturer.		
Check to see that filters are installed correctly.		
Replace damaged filters.		

Provide any suggestions or recommendations for follow-up activities or any other comments related to the inspection and maintenance activities.

6.51: Air-Cooled Condenser Inspection Checklist

AIR-COOLED CONDENSER INSPECTION CHECKLIST

Maintenance Inspector's Name ____________________

Designation and Location of Item Inspected ____________________

____

Instructions: Perform the specified inspection and maintenance tasks. Insert the date when the activity is complete. Make any comments which are pertinent to future maintenance needs.

Item to Be Inspected	Date	Comments
Inspect wiring and electrical controls for loose connections; charred, broken, or wet insulation; evidence of short-circuiting; wrong size fuses; other electrical deficiencies; tighten, correct, repair, or replace as required.		
Check motor for excessive heat and vibration.		
Lubricate electric motor, if applicable. *Do not over lubricate.*		
Lubricate gearbox and/or drive-shaft bearings.		
Check all piping for leaks and tighten loose connections.		
Check fan for bent blades, unbalance, excessive noise, and vibration.		
Inspect mounting brackets, bolts, fan guards; tighten, repair, or replace as required.		
Check and inspect belts and pulleys; align, adjust, or replace as required.		
Inspect for rust and corrosion. Remove rust and corrosion and apply paint where applicable.		
Clean condenser coils and straighten fins as required.		

Provide any suggestions or recommendations for follow-up activities or any other comments related to the inspection and maintenance activities.

6.52: Air Conditioning—Window and Fan Coil Unit Inspection Checklist

AIR CONDITIONING—WINDOW AND FAN COIL UNIT INSPECTION CHECKLIST

Maintenance Inspector's Name ______________________________

Designation and Location of Item Inspected ______________________________

____

Instructions: Perform the specified inspection and maintenance tasks. Insert the date when the activity is complete. Make any comments which are pertinent to future maintenance needs.

Item to Be Inspected	Date	Comments
Replace air filter. Clean cooling/heating coil.		
Lubricate electric motor, as applicable. *Do not over lubricate.*		
Inspect wiring and electrical controls for loose connections; tighten, repair, or replace if necessary.		
Inspect piping and valves for leaks. Tighten or repair as required.		
Check motor(s) for excessive heat and vibration.		
Inspect for rust and corrosion. Remove rust and corrosion and apply paint when applicable.		
Inspect guards and covers. Tighten or adjust as necessary to ensure secure fit.		
Check and ensure that condenser drain line is open.		

Provide any suggestions or recommendations for follow-up activities or any other comments related to the inspection and maintenance activities.

6.53: Air-Conditioning Unit Inspection Checklist

AIR-CONDITIONING UNIT INSPECTION CHECKLIST

Maintenance Inspector's Name ____________________

Designation and Location of Item Inspected ____________________

____

Instructions: Perform the specified inspection and maintenance tasks. Insert the date when the activity is complete. Make any comments which are pertinent to future maintenance needs.

Item to Be Inspected	Date	Comments
Change air filter as required.		
Check condition of coil. Clean as required.		
Check piping for leaks.		
Check compressor operation and valve operation.		
Check and inspect belt; adjust tension as required.		
Lubricate shaft bearings.		
Check wiring and electrical controls for loose connections; charred, frayed, or broken insulation. Check operation of strip heaters. Tighten, repair, or replace if necessary.		
Check for rust and corrosion. Remove rust and corrosion and apply paint where required.		

Provide any suggestions or recommendations for follow-up activities or any other comments related to the inspection and maintenance activities.

6.54: Air Compressor Inspection Checklist

AIR COMPRESSOR INSPECTION CHECKLIST

Maintenance Inspector's Name ______________________________

Designation and Location of Item Inspected ______________________________

____

Instructions: Perform the specified inspection and maintenance tasks. Insert the date when the activity is complete. Make any comments which are pertinent to future maintenance needs.

Item to Be Inspected	Date	Comments
Inspect wiring and electrical controls for loose connections; charred, broken, or wet insulation; short circuits; tighten, repair, or replace as required.		
Clean air intake muffler/filter.		
Clean cylinder fins.		
Test all safety valves.		
Check oil level in crank case; add oil as required.		
Drain condensate from air tank.		
Check and inspect belts and adjust or replace if necessary.		
Check foundation bolts for tightness.		
Clean the trigger valve strainers.		
Check motor for excessive heat and vibration.		
Lubricate electric motor as applicable.		
Check all screws and nuts for tightness.		

Provide any suggestions or recommendations for follow-up activities or any other comments related to the inspection and maintenance activities.

6.55: Aeration Equipment Inspection Checklist

AERATION EQUIPMENT INSPECTION CHECKLIST

Maintenance Inspector's Name ______________________________

Designation and Location of Item Inspected ______________________________

____

Instructions: Perform the specified inspection and maintenance tasks. Insert the date when the activity is complete. Make any comments which are pertinent to future maintenance needs.

Item to Be Inspected	Date	Comments
Check for clogged nozzles, water spread, leakage, rust, corrosion, missing supports, and fence damage. Clean nozzles, adjust, repair, or replace as required.		
Check for algae, dirt, deterioration, and damage to enclosure and screens. Clean, repair, or replace as required.		
Check for proper operation of air diffuser. Drain air diffuser; examine for leaks, breakage, rust, and corrosion. Clean, adjust, repair, or replace as required.		
Inspect wiring and electrical controls for loose connections; charred, broken, or wet insulation; evidence of short-circuiting and other deficiencies. Tighten, repair, or replace as required.		
Lubricate motor as required. *Do not over lubricate.*		
Check drive unit and/or pump for proper operation.		
Check packing glands for leakage. Tighten or replace as required.		
Check for rust and corrosion. Remove rust and corrosion and apply paint where applicable.		
Check valves for proper operation. Lubricate valve stem threads and packing as required.		
Lubricate all fittings.		

Provide any suggestions or recommendations for follow-up activities or any other comments related to the inspection and maintenance activities.

6.56: Dust Control System Check Sheet

DUST CONTROL SYSTEM CHECK SHEET

System Number ____________ Location ____________ Drawing Number ____________

Date Started ______________________ Column Number/Bay ______________________

Equipment Exhausted ____

____

Collection equipment:

Make ____________ Type ____________ Size ____________ Serial ____________

Miscellaneous data ______________________ Design RPM ______________________

Exhauster:

Make ____________ Type ____________ Size ____________ Serial ____________

Shaft diameter ____________ Kilowatt ____________ Design RPM ____________

Design horsepower ____________ Actual horsepower ____________ (at start-up)

Motor and base:

Make ________________ Type ________________ Enclosure ________________

Horsepower _______ RPM _______ Volts _______ Phase _______ Cycle _______

Shaft diameter ______________________ Kilowatt ______________________

V-belt drive:

Motor sheave make __________ No. grooves __________ Belt section __________

Bore ____________________________ Kilowatt ____________________________

Exhauster sheave make _________ No. grooves _________ Belt section _________

Bore ____________________________ Kilowatt ____________________________

System check:

System data checkpoint no.	Design volume (cfm)	Design pressure (in. water gauge)	Balance system (in. water gauge)
1.			
2.			
3.			
etc.			

6.56: *(Continued)*

FORM 6.56

Key Use of Form:

To document general information about the dust-control system and the results of a system controls inspection.

Who Prepares:

Person performing the system check.

Who Uses:

Plant engineering or maintenance department personnel to evaluate efficiency of system and to plan future maintenance and/or repair.

How to Complete:

Provide the information requested in the appropriate spaces.

Alternative Forms:

Changes can be made to this form to be more company-specific.

6.57: Pneumatic Temperature Controls Maintenance Checklist

PNEUMATIC TEMPERATURE CONTROLS MAINTENANCE CHECKLIST

Maintenance Inspector's Name ______________________ Date of Inspection ____________

AIR COMPRESSORS AND ACCESSORIES	(X)
Drain and inspect tank	
Drain and refill crankcase oil	
Oil electric motor	
Clean and/or replace air intake filter	
Inspect tank valve stem packing	
Check belt tension; replace worn belt	
Examine motor brushes and cummutator	
Inspect and/or replace pressure switch contacts	
Check pressure relief valve operation	
Check pressure reducing valve setting	
Check airlines for moisture, oil, and dirt	
Check sediment and water trap	
THERMOSTATS	
Check calibration	
Check throttling range	
Check air connections	
Adjust system sequence	
HYGROSTATS	
Check calibration	
Check throttling range	
Clean or replace element	
Check air connections	
Adjust system sequence	
DAMPERS AND DAMPER MOTORS	
Check travel and close-off	
Lubricate blade and linkage bearings	
Check motor connections and spring	
Check air connections	
Adjust system sequence	

VALVES	(X)
Check operation for tight close-off	
Lubricate valve stem	
Inspect and/or replace packing	
Replace disc where necessary	
Check air connection	
Adjust system sequence	
PRESSURE REGULATORS	
Check calibration	
Check throttling range	
Check pressure setting	
Adjust system sequence	
RELAYS, PILOT VALVES, SWITCHES	
Check operation	
Adjust system sequence	
THERMOMETERS	
Check calibration	

OTHER: ______________________________

COMMENTS: ______________________________

6.57: *(Continued)*

FORM 6.57

Key Use of Form:

To inspect and perform preventative maintenance activities on pneumatic temperature controls.

Who Prepares:

Maintenance personnel performing activities.

Who Uses:

Designated personnel in plant engineering or maintenance department to evaluate efficiency of preventative maintenance program and plan future maintenance activities on controls.

How to Complete:

After performing each activity, place an *X* in the column opposite the task.

Alternative Forms:

A company can revise the form to meet its specific needs.

C. SPECIAL FORMS

There are many maintenance-related activities in every company that must be included in the organization's overall management program. Most of these tasks are unique to the specific company. To assist management, special forms can be developed for the specific activity to control the respective maintenance function more effectively and efficiently.

Every company should evaluate its own operations to ascertain where such forms would help to make the management function more efficient. Once these activities are identified, appropriate documents can be developed. Some examples relating to housekeeping activities are included in this part of the section.

The balance of this section presents some special forms for the control of specific mechanical-related maintenance activities. The Mechanical Downtime Report is used to keep track of loss production time due to failure of mechanical equipment and/or systems. Any overtime required to maintain or repair mechanical equipment is documented for future use on the Mechanical Services Overtime Report. The use and frequency of changing oil and other lubricants need to be monitored for the effective operation of equipment. The Lubrication Schedule and Oil Consumption forms have been provided to document these activities.

6.58: Mechanical Downtime Report

MECHANICAL DOWNTIME REPORT

Date ________________ Shift ________________ Department ________________

Machine	Time Lost Due to Breakdown	Cause of Breakdown	Action Taken

Operator ______________________________

FORM 6.58

Key Use of Form:

To report (document) downtime on mechanical equipment.

Who Prepares:

Operator of machine.

Who Uses:

Supervisor of department to plan and schedule repairs. Also, by personnel in the plant engineering department to evaluate efficiency of machinery and plan and schedule future maintenance-related activities.

How to Complete:

Provide the information requested in the appropriate spaces.

Alternative Forms:

A company can make revisions to the form to customize it for their needs.

6.59: Mechanical Services Overtime Report

MECHANICAL SERVICES OVERTIME REPORT

Person Completing Form ______________________ Date ____________

Times ______________________ Job Number ______________________

Job Location ______________________ Department ______________________

Description of Work to Be Done:

Operations Affected:

Personnel Performing Work:

Summary of Cost Information:

Comments:

6.60: Lubrication Schedule

LUBRICATION SCHEDULE

Week of ____________ Lubricator ____________ Lubricator ____________ Lubricator ____________

Description	No. of Fittings	Lube	Frequency	Midnight							Day							Afternoon						
				M	T	W	T	F	S	S	M	T	W	T	F	S	S	M	T	W	T	F	S	S

COMMENTS:

NOTED SAFETY PROBLEMS:

6.60: *(Continued)*

FORM 6.60

Key Use of Form:

To schedule lubrication of equipment.

Who Prepares:

Scheduling personnel in plant engineering department.

Who Uses:

Management responsible for the lubrication activity.

How to Complete:

Provide the general information requested and note the time the lubrication activity should be performed for each piece of equipment by checking the appropriate boxes.

Alternative Forms:

Can revise this form to make it more company-specific.

6.61: Oil Consumption Form

OIL CONSUMPTION FORM

Equipment Designation ______________ Location ______________

Form Completed By ______________ Date Form Completed ______________

Date	Oiler Initials	Consumption by Oil Type	Remarks

FORM 6.61

Key Use of Form:

To keep track of the consumption of oil by any one piece of mechanical equipment.

Who Prepares:

Supervisor of department in which equipment is located.

Who Uses:

Designated personnel in maintenance department to evaluate oil consumption and to use in planning for equipment maintenance and overhaul.

How to Complete:

Provide the information requested in the appropriate spaces.

Alternative Forms:

Revisions can be made to make form more company-specific.

Section 7

ELECTRICAL EQUIPMENT AND SYSTEMS

This section provides forms that can be used to manage the inventory, inspection, maintenance, and repair of electrical equipment and systems. Included are formats for motors, motor controls, protective devices, batteries, transformers, distribution, and lighting systems.

You should note that some of the forms call for information relating to the mechanical aspects of equipment such as motors. Therefore, there is some overlap between this section and Section 6. For additional formats that can be used in the planning, estimating, scheduling, ordering, and evaluating of electrical maintenance activities, refer to Section 8, "Maintenance Management."

A. INVENTORY FORMS

This section begins with a series of inventory-type forms, referred to as record forms. There should be one form completed for each type of electrical equipment and systems. As inspections, repairs, and other maintenance activities are performed on each item, the date accomplished, along with other pertinent facts such as maintenance activities performed, should be recorded on the form. The person who will complete the form will depend on the type and size of company.

This form is open-ended in that information will continue to be added to it. It serves as the historical basis for all maintenance aspects of the electrical equipment and systems.

7.1: Electrical Equipment Inventory Record

ELECTRICAL EQUIPMENT INVENTORY RECORD

Mfg. Name		Mfg. Order No.	Serial No.
Equipment No.	Type or Model	Drawing No.	Our P.O. No.
Vendor	Approp. No.	Shop Order No.	
Equipment Cost	Installation Cost	Date Installed	Location
Equipment Description			

Maintenance Requirements	Electrical Equipment				
	Equipment				
	Make				
	Serial No.				
	Type Frame				
	Voltage				
Inspection Requirements	Phase				
	H.P.				
	R.P.M.				
	Drive				
	Circuit				
	Date Installed				
Lubrication	Cost				
Manual Page Lube Period					

7.2: Equipment and Repair Record

EQUIPMENT AND REPAIR RECORD

Department		No.
Equipment Location		
Equipment Description		
Manufacturer Name		No.
Serial No.	Purchase Order No.	No.
Date Received	Date of Operation	
Installed by	Work Order No.	
Purchase Price	Installation Cost	
Sold	Transferred	Scrapped
Electrical Equipment: Complete Data		
Mechanical Equipment: Complete Data		
Miscellaneous Equipment		

REPAIR RECORD

Date	Order No.	Reason for Repair	Date	
			Off	On

7.3: Motor Maintenance Record

MOTOR MAINTENANCE RECORD

Plant No. ________ Machine No. ________ Inventory No. ________ Motor No. ________

HP ________ Manufactured By ________

Series	Shunt	Compound	Synchronous	Induction
Type	Frame	Speed	Volts	Amperes
Phase	Cycles	Temp. Rise	Excitation Amps.	Rotor or Armature or Service No.
Model No.	Form No.	Manufacturer's No.	Serial No.	
Mfgr's. Order No.	Our Order No.	Date		
Connection Diagram–Rotor or Armature	Stator			

SPECIFICATION	BEARINGS	SHAFT EXTENSION	PULLEY	GEAR	V-BELT DRIVE
Open ☐ Exp. Proof ☐ Drip Proof ☐ Totally Encl. ☐ Vertical ☐ ________ ☐	Sleeve ☐ Ball ☐ Roller ☐	Dia. ________ Length ________ Keyway ________	Dia. ________ Face ________	Teeth ________ Pitch ________ Face ________	No. Grooves ________ Pitch Dia. ________ Belt Section ☐ A—$1/2 \times 11/32$ ☐ B—$21/32 \times 7/16$ ☐ C—$7/8 \times 17/32$ ☐ D—$1\,1/4 \times 3/4$

Motor Service Record

Date Installed	Location	Application

Date Repaired	Repairs or Parts Replaced	Cause	Repaired By	Total Cost

7.3: *(Continued)*

Name of Part	No. Per Motor	Manufacturer's No.	Parts Purchased								
			Date	Quan.	Tot. Cost	Date	Quan.	Tot. Cost	Date	Quan.	Tot. Cost
Rotor or Armature Coils											
Stator Coils											
Field Coils—Shunt											
Field Coils—Series											
Field Coils—Commutating											
Assembled Seg. or Coll. Rings											
Brushes											
Brushholders											
Brushholder Springs											
Brushholder Fingers											
Bearing—Front											
Bearing—Rear											
Oil Ring—Front											
Oil Ring—Rear											

STARTER

Class	HP	Style or Stock Order	Remarks

7.4: Motor, Generator, and Control Services Record

MOTOR, GENERATOR, AND CONTROL SERVICES RECORD

Motor HP	A-C	D-C	Type	Frame	R.P.M.	Phase	Cycles	Volts	Division	Section
Winding	Tool No.			Manufacturer's No.			Serial No.		Location-Section	Col. No.
Date Installed	Date in Storage			Date Installed		Date in Storage		Date Installed	Date in Storage	
Drives Machine			Tool No.	Drives Machine			Tool No.	Drives Machine		Tool No.

Date	Blownout	Bearings	End Play	Brushes & Holder	Commu.	Megger	General Condition	Inspected By

Date Taken Out of Service ______________________ Date Scrapped ______________________

7.4: *(Continued)*

Disconnect Switch Manufacturer's No.	Starter Switch or Controller Style No.	Serial No.	Push Button Style No.	Limit Switch Style No.	Date Installed

Date	Check Connections	Clean	Check Contacts	Check for Worn Parts	Check Spring Tension and Contact Pressure	Check Protective Device	Check Circuit Voltage	Inspected By

B. INSPECTION FORMS

The next group of forms found in this section are inspection/service checklists. They are used in the inspection or service activities for the specific piece of electrical equipment.

The person making the inspection should first record the general types of information requested such as Date of Inspection, Item Being Inspected, and so on. The next step is the inspection and recording of observations on this document along with any suggested follow-up activities that should take place such as an inspection, a need for immediate repair, or a recommendation for replacement. More than one day may be needed to complete the inspection process.

After the form is completed, it should be transmitted to the designated maintenance personnel. In addition, copies should be kept on file for future reference.

7.5: Electrical System (Buildings) Inspection Checklist

ELECTRICAL SYSTEM (BUILDINGS) INSPECTION CHECKLIST

Maintenance Inspector's Name ______________________________

Description and Location of Items Being Inspected ______________________________

____

Instructions: Perform the specified inspection and maintenance tasks. Insert the date when the activity is complete. Make any comments which are pertinent to future maintenance needs.

Item to Be Inspected	Date	Comments
Inspect wiring and electrical controls for loose connections; charred, broken, or wet insulation; evidence of short-circuiting and other deficiencies. Tighten, repair, or replace as required.		
Check for damaged wiring devices, defective insulators, cleats and cable supports, or exposed live parts.		
Check cable sag, cable spacing, number of conductors in conduit and raceway.		
Check switches and breakers for condition, alignment of contacts, and signs of arcing.		
Check convenience outlets for condition, dirt, and signs of arcing.		

Provide any suggestions or recommendations for follow-up activities or any other comments related to the inspection and maintenance activities.

7.6: Lighting System Inspection Checklist

LIGHTING SYSTEM INSPECTION CHECKLIST

Maintenance Inspector's Name ______________________________

Date of Inspection ____________________ Location ____________________

Item Being Inspected ______________________________

LIGHTING FIXTURES

_____ Inadequately supported, insecure, and improperly located; evidence of unauthorized removal and relocation.

_____ Incorrect types installed in hazardous locations; change in facility class requires replacement.

_____ Improperly located in clothes closets. (Should be above door or in ceiling and not serviced with cord pendants.)

_____ Cracked or broken luminaires and fixture parts, missing pullcords, metal pull-chains not provided with insulating links.

_____ Indications of objects being supported from, hung on, or stored in fixtures.

_____ Evidence of overheating, undersized, or other damage to sockets, exposed or damaged connecting wiring.

_____ Repair or replace exposed live wires, undersized or overheated sockets, broken or missing pullcords, cracked glass luminaires, and missing fixture parts.

LAMPS

_____ Oversized, blistering, loose base, thermal cracks from contact with fixture, bare lamps in hazardous locations, poor socket lamp connections, improper types for special applications.

_____ Operation of fluorescent fixtures shows poor burning and starting characteristics, and loud humming ballasts.

SWITCHES

_____ Defective operation, broken or missing parts, arcing noises.

CONVENIENCE OUTLETS

_____ Dirty, inadequate, defective contacts, difficult plugging, overheating, evidence of overloading on multiple sockets servicing lamps or appliances, lack of grounding terminal. Revamp local wiring to eliminate dangerous and improper conditions.

CORDS, CORD EXTENSIONS, AND PORTABLE AND APPLIANCE CORDS

_____ Inadequate, unsafe, unreliable, incorrect types being used.

_____ Lengths too long, poor insulation, twisted, spliced, exposed to damage underfoot, lying on floor or across heated surfaces or lamps; lamp types used for portable extensions that are subject to moisture, oil, and grease.

7.6: *(Continued)*

____ Plugs: cracks, breaks, loose connections, wires improperly attached and in danger of pulling away from plug when removing from outlet, missing protective cover on male ends, no grounding terminal or ground wire with clamp.

LIGHTING VOLTAGE

____ Spot measurement at fixture and lighting outlets indicates measured voltage in excess of 5 percent of nominal lamp voltages.

____ Unauthorized connections of hot plates, coffee pots, heating devices, and other electrical equipment on lighting circuits.

____ Interference with branch circuits for power and lighting from motor starting or stopping, such as excessive light flicker, or excessive voltage dips causing fluorescent and mercury lamps to drop out.

ILLUMINATION LEVELS

____ Ambient conditions such as dirty walls and ceilings.

____ Spot measurement of light levels using accurate foot candle meter indicates depreciation of 20 to 25 percent of level obtainable from clean fixtures and new lamps seasoned for 100 hours.

____ Failure to keep continuous record of illumination levels at established checkpoints.

COMMENTS:

7.7: Distribution Transformer Inspection Checklist

DISTRIBUTION TRANSFORMER INSPECTION CHECKLIST

Maintenance Inspector's Name ______________________________

Date of Inspection ______________________ Location ______________________

Item Being Inspected ______________________________

DEENERGIZED STATE:

BUSHINGS AND INSULATORS

(Remove all grease, dirt, and other foreign materials by washing and then drying.)

____ Insulators and Porcelain Parts: indications of cracks, checks, chips, breaks; where flashover streaks are visible, re-examine for injury to glaze or for presence of cracks.

____ Chipped glaze exceeding $1/2$ inch in depth or an area exceeding one square inch on any insulator or insulator unit; report for investigation by a qualified electrical engineer.

____ Severe cracks, chipped cement, or indications of leakage around bases of joints of metal to porcelain parts at terminal and transformer ends.

____ Terminal Ends: mechanical deficiencies, looseness, corrosion, damage to cable clamps.

____ Improper oil level in oil-filled bushings (fill bushing if oil is below proper level).

____ Heating evidenced by discolorations, and corrosion indicated by blue, green, white, or brown corrosion products on metallic portions of all main and ground terminals, including terminal board and grounding connections inside transformer case. (Clean metalwork, disconnecting if required, and cover with thin coating of nonoxide grease; if connections are disassembled, rough spots on contact surfaces should be filed smooth, and all projections removed; see that all bolted and crimped connections are tight by setting up nuts or recrimping when looseness is evident or suspected; clean and tighten all corroded or loose connections; repair or replace cables with frayed and broken strands; repair frayed or broken cable insulation.)

ENCLOSURE AND CASES

____ If case is opened for any reason, examine immediately for signs of moisture inside cover, and where present, for plugged breathers, inactive desiccant, enclosure leakage. (Protect transformer liquid from dust, dirt, and windblown debris by covering open tank with temporary cover made of wood, kraft paper, plastic sheeting, or other suitable dust tight material; clear plugged openings; if desiccant is inactive, replace with fresh material or reactivate for proper functioning; if rust or corrosion is evident on inside cover, clean and paint with preservative.)

COILS AND CORES

____ When cover is open, examine interior for deficiencies, dirt, sludge. If feasible, probe down sides with glass rod, and if dirt and sludge exceeds approximately $1/2$ inch, arrange to change or filter insulating oil, and have coils and cores cleaned. (Use low-pressure air, if available, to blow out dust from air-cooled transformers, or pull out dust with vacuum equipment.)

7.7: *(Continued)*

GAUGES AND ALARMS

- ____ Liquid Level Gauge and Alarm System: dirty, not readable, improper frequency of calibration. Engineering tests should be performed under supervision of qualified electrical engineer before, during, or after inspection, as applicable. Assistance of inspectors and craft personnel is required, and arrangements should be made with proper authority to assure coordinated effort by everyone taking part.
- ____ Test grounding system.
- ____ Measure load current with recording meter over period of time when load is likely to be at its peak; measure peak-load voltage; make regulation tests and tests of operating temperature during peak-load-current tests; test and calibrate thermometers or other temperature alarm systems.
- ____ Test dielectric strength of insulating liquid.
- ____ Test insulation resistance.

ENERGIZED STATE:

CONCRETE FOUNDATIONS AND SUPPORTING PADS

- ____ Settling and movement, surface cracks exceeding 1/16 inch in width, breaking or crumbling within 2 inches of anchor bolts.
- ____ Anchor Bolts: loose or missing parts, corrosion, particularly at points closest to metal base plates and concrete foundations resulting from moisture or foreign matter, and exceeding 1/8 inch in depth.

MOUNTING PLATFORMS, WOODEN

- ____ Cracks, breaks, signs of weakening around supporting members; rot, particularly at bolts and other fastenings, holes through which bolts pass, wood contacting metal.
- ____ Burning and charring at contact points, indicating grounding deficiency.
- ____ Inadequate wood preservation treatment.

MOUNTING PLATFORMS, METALLIC

- ____ Deep pits from rust, corrosion; other signs of deterioration likely to weaken structure.

HANGERS, BRACKETS, BRACES, AND CONNECTIONS

- ____ Rust, corrosion, bent, distorted, loose, missing, broken, split, other damage; burning or charring at wood contact points caused by grounding deficiency.

ENCLOSURES, CASES, AND ATTACHED APPURTENANCES

- ____ Collections of dirt or other debris close to enclosure that may interfere with radiation of heat from transformer or flashover.
- ____ Dirt, particularly around insulators, bushings, or cable entrance boxes.
- ____ Leaks of liquid-filled transformers.
- ____ Deteriorated paint, scaling, rust; corrosion, particularly at all attached appurtenances, such as lifting lugs, bracket connections, and metallic parts in contact with each other.

7.7: *(Continued)*

NAMEPLATES AND WARNING SIGNS

_____ Dirty, chipped, worn, corroded, illegible, improperly placed.

GROUNDING

_____ Visual Connections: loose, missing, broken connections; signs of burning or overheating, corrosion, rust, frayed cable strands, more than 1 strand broken in 7-strand cable, more than 3 strands in 19-strand cable.

BUSHINGS AND INSULATORS

_____ Cracked, chipped, or broken porcelain; indication of carbon deposits; streaks from flashovers; dirt, dust, grease, soot, or other foreign material on porcelain parts; signs of oil or moisture at point of insulator entrance.

GROUNDING AND PHASE TERMINALS

_____ Overheating evidenced by excessive discolorations of cooper, loose connection bolts, defective cable insulation; no mechanical tension during temperature changes; leads appear improperly trained and create danger of flashovers from unsafe phase-to-phase or phase-to-ground clearances caused by deterioration of leads or expansions during temperature changes.

BREATHERS

_____ Holes plugged with debris; desiccant-type breathers need servicing or replacement.

GRILLS AND LOUVERS FOR VENTILATION OF AIR-COOLED TRANSFORMERS

_____ Plugged with debris or foreign matter interfering with free passage of air. (Openings located near floor or ground line can be inspected with small nonmetallic framed mirror having long insulated handle, used in conjunction with light from hand flashlamp having insulated casing. Throw light beam onto mirror and reflect upward into openings.)

COMMENTS:

7.8: Power Transformer Inspection Checklist

POWER TRANSFORMER INSPECTION CHECKLIST

Maintenance Inspector's Name ______________________________

Date of Inspection ____________________ Location ____________________

Item Being Inspected ______________________________

DEENERGIZED STATE:

BUSHINGS AND INSULATORS

(Remove all grease, dirt, and other foreign materials by washing and then drying.)

____ Insulators and Porcelain Parts: indications of cracks, checks, chips, breaks; where flashover streaks are visible, re-examine for injury to glaze or for presence of cracks.

____ Chipped glaze exceeding 1/2 inch in depth or an area exceeding one square inch on any insulator or insulator unit; report for investigation by a qualified electrical engineer.

____ Severe cracks, chipped cement, or indications of leakage around bases of joints of metal to porcelain parts at terminal and transformer ends.

____ Terminal Ends: mechanical deficiencies, looseness, corrosion, damage to cable clamps.

____ Improper oil level in oil-filled bushings (fill bushing if oil is below proper level).

____ Heating evidenced by discolorations, and corrosion indicated by blue, green, white, or brown corrosion products on metallic portions of all main and ground terminals, including terminal board and grounding connections inside transformer case.

____ Pipe, Bar Copper, and Connections: indications of overheating or flashover fusing.

____ Cable Connections: broken, burned, corroded, missing strands. (Fused portions of connectors, cables, pipe, or bus copper should be filed smooth and all projections removed; clean metalwork, disconnecting if required and cover with thin coating of nonoxide grease; if connections are disassembled, rough spots on contact surfaces should be filed smooth and all projections removed; see that all bolted and crimped connections are tight by setting up nuts or recrimping when looseness is evident or suspected; clean and tighten all corroded or loose connections; repair or replace cables with frayed and broken strands; repair frayed or broken cable insulation.)

ENCLOSURE AND CASES

____ If case is opened for any reason, examine immediately for signs of moisture inside cover, and where present, for plugged breathers, inactive desiccant, enclosure leakage. (Protect transformer liquid from dust, dirt, and windblown debris by covering open tank with temporary cover made of wood, kraft paper, plastic sheeting, or other suitable dust tight material; clear plugged openings; if desiccant is inactive, replace with fresh material or reactivate for proper functioning; if rust or corrosion is evident on inside cover, clean and paint with preservative.)

7.8: *(Continued)*

COILS AND CORES

_____ When cover is open, examine interior for deficiencies, dirt, sludge. If feasible, probe down sides with glass rod, and if dirt and sludge exceed approximately $^1/_2$ inch, arrange to change or filter insulating oil, and have coils and cores cleaned.

BUSHING-TYPE INSTRUMENT TRANSFORMERS

_____ Indications of deteriorated insulation: overheating evidenced by excessive discoloration of terminals and other visible copper; physical strains indicated by bent or distorted members.

_____ Terminals, Including Secondaries: corrosion, loose connections.

_____ Secondary Leads: visible broken, cracked, or frayed insulation.

_____ Conduit and Associated Fittings Carrying Secondary Leads: rust, corrosion, other deterioration, loose joints in conduit fittings and around terminal boxes. (Clean, tighten, or repair terminals; tighten all loose or defective conduit-supporting clamps; clean and paint conduit and associated fitting areas showing rust and corrosion; if fuse box for bushing-type potential transformers is installed, check fuses for proper size, as specified by manufacturer or station engineers; assure that fuses have not been shorted out or bridged; replace blown or improperly sized fuses.)

AUTOMATIC TAP-CHANGERS (load ratio control apparatus)

_____ Make inspection in accordance with manufacturer's instructions. (Clean and lubricate all moving parts and contacts in accordance with manufacturer's recommendations.)

FORCED-AIR FANS AND FAN CONTROLS

_____ Fans and Motors: defective bearings, inadequate lubrication, presence of dirt, bent or broken fan blades or guards, lack of rigidity of mountings, indications of corrosion or rust. (Make minor repairs as necessary to assure dependable and continuous service until next inspection.)

_____ Starting and Stopping Devices: improper functioning as determined from operating once or twice.

_____ Fan Speed: not in accordance with nameplate requirements.

WATER COOLING SYSTEMS

_____ Water not being delivered in required quantity.

GAUGES AND ALARMS

_____ Liquid Level Gauge and Alarm System: dirty, not readable, improper frequency of calibration.

_____ Pressure Gauges and Valves on Inert Gas Systems: improper frequency of gauge calibration; leaks in piping both before opening and after closing tanks; apply soap bubble test to joints and connections when pressure is unsteady. Engineering tests should be performed under supervision of qualified electrical engineer before, during, or after inspection, as applicable. Assistance of inspectors and craft personnel is required, and arrangements should be made with proper authority to assure coordinated effort by everyone taking part.

7.8: *(Continued)*

_____ Test grounding system.

_____ Measure load current with recording meter over period of time when load is likely to be at its peak; measure peak-load voltage; make regulation tests and tests of operating temperature during peak-load current tests; test and calibrate thermometers or other temperature alarm systems.

_____ Test dielectric strength of insulating liquid.

_____ Test insulation resistance.

ENERGIZED STATE:

CONCRETE FOUNDATIONS AND SUPPORTING PADS

_____ Settling and movement, surfacc cracks exceeding 1/16 inch in width, breaking or crumbling within 2 inches of anchor bolts.

_____ Anchor Bolts: loose or missing parts, corrosion, particularly at points closest to metal base plates and concrete foundations resulting from moisture or foreign matter, and exceeding 1/8 inch in depth.

MOUNTING PLATFORMS, WOODEN

_____ Cracks, breaks, signs of weakening around supporting members; rot, particularly at bolts and other fastenings, holes through which bolts pass, wood contacting metal.

_____ Burning and charring at contact points, indicating grounding deficiency.

_____ Inadequate wood preservation treatment.

MOUNTING PLATFORMS, METALLIC

_____ Deep pits from rust, corrosion; other signs of deterioration likely to weaken structure.

HANGERS, BRACKETS, BRACES, AND CONNECTIONS

_____ Rust, corrosion, bent, distorted, loose, missing, broken, split, other damage; burning or charring at wood contact points caused by grounding deficiency.

ENCLOSURES, CASES, AND ATTACHED APPURTENANCES

_____ Collections of dirt or other debris close to enclosure that may interfere with radiation of heat from transformer or flashover.

_____ Dirt, particularly around insulators, bushings, or cable entrance boxes.

_____ Leaks of liquid-filled transformers.

_____ Deteriorated paint, scaling, rust; corrosion, particularly at all attached appurtenances, such as lifting lugs, bracket connections, and metallic parts in contact with each other.

NAMEPLATES AND WARNING SIGNS

_____ Dirty, chipped, worn, corroded, illegible, improperly placed.

GASKETS

_____ Leakage, cracks, breaks, brittleness.

7.8: *(Continued)*

INERT GAS SYSTEMS

_____ Incorrect pressure in system. (Maximum: 3 to 5 pounds, Minimum $^1/_4$ to 1 pound.)

_____ Pipe and Valve Connections: leaking gas (indicated by liquid oozing out of joints or by soapsuds test).

_____ Loose gas tank fastenings, loose valves.

_____ If previous arrangements were made with operating forces, bleed a little gas from system by means of pressure-regulating device; note evidence of leaks.

BUSHINGS AND INSULATORS

_____ Cracked, chipped, or broken porcelain; indication of carbon deposits; streaks from flashovers; dirt, dust, grease, soot, or other foreign material on porcelain parts; signs of oil or moisture at point of insulator entrance.

GROUNDING AND PHASE TERMINALS

_____ Overheating evidenced by excessive discolorations of cooper, loose connection bolts, defective cable insulation; no mechanical tension during temperature changes; leads appear improperly trained and create danger of flashovers from unsafe phase-to-phase or phase-to-ground clearances caused by deterioration of leads or expansions during temperature changes.

INSTRUMENT TRANSFORMER JUNCTION BOXES AND CONDUITS

_____ Loose or severely corroded components, including secondary lead connections.

BREATHERS

_____ Holes plugged with debris; desiccant-type breathers need servicing or replacement.

TEMPERATURE-INDICATING AND ALARM SYSTEMS, INCLUDING CONDUIT AND FITTINGS

_____ Loose fastenings, rust, severe corrosion, deteriorated paint, other mechanical defects, loose electrical connections.

MANUAL AND AUTOMATIC TAP CHANGERS

_____ Loose connections, rust, severe corrosion, other mechanical defects, lack of lubrication, signs of burning around conducting and nonconducting parts of terminal boards.

LIQUID LEVEL INDICATORS

_____ Rust, corrosion, lack of protective paint; cracked or dirty gauge glasses so that liquid level not discernible; plugged gauge-glass piping, liquid level below permissible level indicated by mark for gauging; signs of leakage around piping, gauge cocks, gauge glasses, or other indicating devices.

FANS AND FAN CONTROLS FOR AIR-COOLED TRANSFORMERS

_____ Lack of rigidity in mounting fastenings.

_____ Motors (external): dirty, moisture, grease, oil, overheating, detrimental ambient conditions.

7.8: *(Continued)*

_____ Apparent deterioration of open wiring and conduit that may cause malfunctioning of either fans or controls.

_____ Improper functioning when manual (not automatic) fan controls operated.

WATER COOLING SYSTEMS

_____ Leaks in piping, fittings, or valve; visible drainage system plugged; open ditches for drainage water fouled with vegetation.

_____ Bearings: evidence of wear, indications of corrosion, external deterioration, leaks.

_____ Incipient deterioration, corrosion, rust, loose fastenings, other mechanical deficiencies, loose electrical connections for all components of alarm system.

_____ Temperature Devices: signs of deterioration that might cause malfunction or difficulty in taking readings.

_____ When pressure gauge readings on each side of strainer varies more than a pound or two, look for cause, such as plugged strainer.

GROUNDING

_____ Visual Connections: loose, missing, broken connections; signs of burning or overheating, corrosion, rust, frayed cable strands, more than 1 strand broken in 7-strand cable, more than 3 strands broken in 19-strand cable.

COMMENTS:

7.9: Electrical Ground Inspection Checklist

ELECTRICAL GROUND INSPECTION CHECKLIST

Maintenance Inspector's Name ______________________________

Date of Inspection ____________________ Location ____________________

Item Being Inspected ______________________________

____ Visual Connections: loose, missing, broken connections; signs of burning or overheating, corrosion, rust, frayed cable strands, more than 1 strand broken in 7-strand cable, more than 3 strands broken in 19-strand cable.

____ Underground Connections: unsatisfactory condition or defects uncovered when 4 or 5 connections are exposed to view by digging.

____ Test Measurement: Permissive resistance

From	To	Permissive Resistance
Point of connection on structure, equipment enclosure, or neutral conductor	Top of ground rod	
Ground rod, mat, or network	Ground (earth)	
Gates	Gateposts	1/2 ohm
Operating rods and handles of group-operated switches	Supporting structure	1/2 ohm
Metallic-cable sheathing	Ground rod, cable, or metal structure	1/2 ohm
Equipment served by rigid conduit	Nearest grounding cable attachment on conduit runs of less than 25 ft.	10 ohms

____ When total resistance in check point 4 or 5 exceeds allowable, measure resistances of individual portions of the circuits to determine the points of excessive resistance and report.

____ Substandard resistance values resulting from poor contact between metallic portions of grounding system and earth.

____ Structural steel, piping, or conduit run exceeding 25 ft. used as a current-carrying part of grounding circuit for protection of equipment.

____ Absence of ground-cable connections.

COMMENTS:

7.10: Motor Inspection Checklist

MOTOR INSPECTION CHECKLIST

Maintenance Inspector's Name ______________________________

Date of Inspection ____________________ Location ____________________

Item Being Inspected ______________________________

GENERAL

____ When practicable, start, run, and cycle motor and generator equipment through load range. Take care in starting motors and generators. On standby or infrequently operated equipment, check rotor freedom and lubrication. At humid locations, check records for evidence of regular exercise; if not found, arrange for drying out windings; megger windings before starting motor.

RUNNING INSPECTION (while equipment operates)

____ Log or Operator Records: evidence of motor or generator overload, low-power factor of load, excessive variations in bearing temperature, operating difficulties.

____ Exposure: unsafe accessibility for maintenance of instrumentation; exposed to physical or other damage from normal plant functions, processes, traffic, and radiant heat; inadequate personnel guards, fences; insufficient, missing, or illegible signs, identification, or operating instructions.

____ Housekeeping: dust, dirt, airborne grit, sand; dripping oil, water, other fluids, vapors; rust, corrosion; peeling, scratches, abrasions or other damage to painted surfaces. Remove oil and solvent cans, oil or solvent soaked rags and waste, other combustibles, particularly those near commutating machinery; remove obstructions that may interfere with rotation or ventilation.

____ Machine Operation: noisy, unbalanced, rubbing, excessive vibration, rattling parts.

____ Structural Supports: inadequate, cracks, settlement; defective or inadequate vibration pads, shockmounts, dampers; loose, dirty, corroded bolts and fittings.

____ Ventilation: dirty, inadequate amount of air passing through machine; dirty, clogged, stator-iron air slots causing excessive temperature. (Too hot to touch. Measured temperature should not exceed 80°C for open frames, or 90°C for enclosed frames. Compare with manufacturer's data.)

____ Motor and Generator Leads: exposed bare conductors; frayed, cracked, peeled insulation; poor taping; moisture, paint, oil, grease; vibration, abrasions, breaks in insulation at entrance to conduit or machines; arcs, burns, overheated, inadequate terminal connections; lack of resiliency, lack of life, dried-out insulation; exposure to physical damage, traffic, water, heat, for semipermanent, temporary, or emergency connections.

____ Bearings: improper lubrication (check lubrication schedules for lubricant used and frequency), improper oil level in oil gauges, incorrectly: reading gauges, noisy bearings, overheated bearing caps or housings. (If bearings are too hot to touch, determine causes. A slow but continuous rise in bearing temperature after greasing indicates possible overlubrication or underlubrication, improper lubricant, or deteriorated bearings. Under normal conditions, the temperature of ball or roller bearings will vary from 10°F to 60°F above the ambient temperature.)

7.10: *(Continued)*

_____ Collector Rings, Commutators, Brushes: excessive sparking, surface dirt, grease (check cleanliness with clean canvas paddle); sparking or excessive brush movement caused by eccentricity, sprung shaft, worn bearings, high bars of mica, surface scratches, roughness; end-play resulting from magnetic-center hunting of rotor; inadequate brush freedom; nonuniform brush wear; poor commutation, improper brushes, incorrect brush pressures. Adjust brush spring pressure to between $1^3/_4$ to $2^1/_2$ psi of brush-commutator contact area for light metallized carbon or graphite brushes; for pressure for other type brushes, check manufacturer's data. (Measure with spring scale.)

_____ Starters, Motor Controllers, Rheostats, Associated Switches: damaged or defective insulation, loose laminations, defective heater or resistance elements, worn contacts, shorts between contacts, arcing, grounds, loose connections, burned or corroded contacts. Replace worn contacts and defective heater or resistance elements.

_____ Protective Equipment: dirty, signs of arcing, symptoms of faulty operation, improper condition of contacts, burned-out pilot lamps, burned-out fuses.

SHUTDOWN INSPECTION (while equipment is not in operation and is electrically disconnected. A shutdown inspection includes a running inspection.)

_____ Stators: dirt, debris, grease; coils not firmly set in slots; burns, tears, aging, embrittlement, moisture in insulation; clogged air slots; rubbing, corrosion, loose laminations of stator-iron; charred or broken slot wedges; abrasion of insulation or chafing in slots; signs of arcing or grounds.

_____ Rotors: difficult turning, rubbing, excessive bearing friction, end play, overheating, looseness of windings, charred wedges, broken, cracked, loosely welded or soldered rotor bars or joints; cracked end rings in squirrel cage motors; loose field spools and deteriorated leads and connections in synchronous motors; deteriorated insulation in wound rotors.

_____ Rotor-Stator Gaps: Check gaps on 5hp or larger induction motors, particularly of the sleeve-bearing type. Where practical, measure and record gaps on the load, pulley, or gear end of the motor. Measure at 2 rotor positions, 180° apart, 4 points for each rotor position. If there is more than 10% variation in gaps, arrange for realignment.

_____ Mechanical Parts: corrosion, improper lubrication, misalignment, end play, interference, inadequate chain or belt tension.

_____ Insulation Resistance: Test insulation resistance of motor and generator windings. Compare results with maintenance information. Insulation resistance values are arbitrary and should be correlated with operating conditions, exposure to moisture, metallic dust, age, length of time in service, severity of service, and maintenance levels.

COMMENTS:

7.11: Motor/Generator Set Inspection Checklist

MOTOR/GENERATOR INSPECTION CHECKLIST

Maintenance Inspector's Name ______________________________

Description and Location of Items Being Inspected ______________________________

____

Instructions: Perform the specified inspection and maintenance tasks. Insert the date when the activity is complete. Make any comments which are pertinent to future maintenance needs.

Item to Be Inspected	Date	Comments
Inspect wiring and electrical controls for loose connections; charring, broken, or wet insulation; evidence of short circuiting and other deficiencies. Tighten, repair, or replace as required.		
Lubricate motor and generator as required. *Do not over lubricate.*		
Operate motor/generator when practicable and check for excessive heat and vibration.		
Check mounting bolts, guards, and brackets as applicable. Tighten, replace, or install as required.		
Inspect for rust and corrosion. Remove rust and corrosion and apply paint where required.		

Provide any suggestions or recommendations for follow-up activities or any other comments related to the inspection and maintenance activities.

7.12: Motor/Pump Inspection Checklist

MOTOR/PUMP INSPECTION CHECKLIST

Maintenance Inspector's Name ______________________________

Description and Location of Items Being Inspected ______________________________

____

Instructions: Perform the specified inspection and maintenance tasks. Insert the date when the activity is complete. Make any comments which are pertinent to future maintenance needs.

Item to Be Inspected	Date	Comments
Check for leaks.		
Check gauge glass.		
Check for leaks around pump-packing gland. Repack, replace, or tighten as required.		
Lubricate pump.		
Check oil level in reduction unit. Add oil as required.		
Lubricate electric motor where applicable. *Do not over lubricate.*		
Check relief valves for proper operation and pressure release. Adjust as required.		
Check motor for excessive noise or vibration.		
Clean filter and strainer.		
Check all pipe hangers and supports, tighten if necessary.		
Check for rust and corrosion. Remove rust and corrosion and apply paint where applicable.		
Inspect wiring and electrical controls for loose connections; charring, broken, or wet insulation; evidence of short-circuiting, and other deficiencies. Tighten, repair, or replace as required.		

Provide any suggestions or recommendations for follow-up activities or any other comments related to the inspection and maintenance activities.

7.13: Motor/Pump Assembly (Hydraulic) Inspection Checklist

MOTOR/PUMP ASSEMBLY (HYDRAULIC) INSPECTION CHECKLIST

Maintenance Inspector's Name ______________________________

Description and Location of Items Being Inspected ______________________________

____

Instructions: Perform the specified inspection and maintenance tasks. Insert the date when the activity is complete. Make any comments which are pertinent to future maintenance needs.

Item to Be Inspected	Date	Comments
Check main relief valve and adjust to maintain proper operating pressure.		
Check relief valve for sticking. Repair or replace as required.		
Check circulation by observing fluid in reservoir.		
Check temperature of oil for proper operation of heat exchanger.		
Check unit for excessive noise and vibration.		
Check pump volume control cylinder for freedom of operation.		
Check oil level in reservoir and add oil if necessary.		
Check complete system for leaks. Repair or replace as required.		
Visually inspect all electrical equipment for deterioration, dust, and moisture.		
Check all shafts couplings for wear and alignment.		
Check motor bearing for excessive heating or unusual noise.		
Lubricate electric motor where applicable. *Do not over lubricate.*		
Check all pipe hangers and supports. Tighten where necessary.		

7.13: *(Continued)*

Item to Be Inspected	Date	Comments
Check motor starter and control relay contacts. Replace where necessary.		
Inspect wiring and electrical controls for loose connections; charring, broken, or wet insulation; evidence of short-circuiting and other deficiencies. Tighten, repair, or replace as required.		
Check pump for proper operation.		

Provide any suggestions or recommendations for follow-up activities or any other comments related to the inspection and maintenance activities.

7.14: Elevator Inspection Checklist

ELEVATOR INSPECTION CHECKLIST

Maintenance Inspector's Name ______________________________

Elevator Designation and Location ____________ Date of Inspection ________

Instructions: Check off items after inspecting each one.

Hoistway

____ Mechanical equipment, items intact and securely fastened to their mountings.

- ____ Sheaves
- ____ Buffers
- ____ Door closers
- ____ Floor selectors
- ____ Limit switch
- ____ Hoistway door hangers
- ____ Door gibs

____ Interlocks are mechanically fastened to their base or mountings and the latching head is securely locked when the door is properly closed.

____ Check hoist and governor ropes and their fastenings for wear and rust.

____ Check traveling cables, make sure they are properly hung and the outer wrapping around the electrical wires is not worn so a short or ground might possibly occur should the traveling cables come in contact with other mechanical equipment in the hoistway.

____ Check rails for proper alignment and tightness of rail fastenings, brackets, and fish plates.

____ Check steadying plates to see that the car is securely fastened to the car frame.

____ Check ropes for vibration and lay.

____ Check guide shoes for excessive float.

____ Check gibs for wear.

____ Check door operators for alignment.

Pit

____ Check oil level in the buffer.

____ Check rope stretch.

____ Check for debris or water leaks.

____ Check safety shoes from the bottom of the car.

7.14: *(Continued)*

Machine Room

_____ Check motors and generators to see that commutators are clean, properly undercut, and equipped with the correct brushes. Brush holders must be clean and brushes well seated, with proper spring pressure, the commutator free from flat spots, high bars, and pitting.

_____ Check bearing wear to determine if it is affecting the armature air gap and rotor clearances.

_____ Check all electrical equipment to see that it is grounded through ground clamps, and machinery should operate without overheating.

_____ Check brakes to make sure that each will hold 125 percent of full load and that lifting application is unimpaired by frozen or worn pins.

_____ Check brake linings to make sure they are securely fastened to the shoe and in good contact with the brake pulley.

_____ Check shafts into the pulley for proper alignment and secure fastenings to the brake pulley.

_____ Check gears and bolts to make sure they are not loose or broken.

_____ Check worm and gear for backlash and end thrust.

_____ Check sheave grooves for uneven wear or bottomed ropes.

_____ Check gland packing on a geared machine.

_____ Check gear-case oil.

_____ Check machine fastenings.

_____ Check controllers for correct fuse capacity, broken leads, loose connections in lugs, loose or broken resistances, improper contacts, worn contacts, weak springs, improper contact-spring tension, proper contact wipe, and worn pins or brushes.

_____ Check switches for residual magnetism and gummy cores.

_____ Check safety equipment for blocked or shorted contacts.

_____ Check governors; be sure the rope is well seated in the sheave, that the operating mechanism is well lubricated and free to move, and that no pins are stiff from rust or paint.

_____ Check landing equipment for broken buttons and lamp fixtures. Emergency key glasses and keys must be in place, and lighting at landings must be bright enough so that people leaving an elevator enter well-lighted areas.

COMMENTS:

7.15: Panel Board Inspection Checklist

PANEL BOARD INSPECTION CHECKLIST

Maintenance Inspector's Name ______________________________

Description and Location of Items Being Inspected ______________________________

____

Instructions: Perform the specified inspection and maintenance tasks. Insert the date when the activity is complete. Make any comments which are pertinent to future maintenance needs.

Item to Be Inspected	Date	Comments
Safety—comply with all current safety precautions.		
Inspect for signs of overheating and corrosion.		
Inspect to ensure that feeder schedules, circuit diagrams, identification charts, etc., are properly posted.		
Inspect for loose or inadequate connections; tighten or repair if necessary.		
Inspect for rust and corrosion. Remove rust and corrosion and apply paint where necessary.		

Provide any suggestions or recommendations for follow-up activities or any other comments related to the inspection and maintenance activities.

7.16: Vault and Manhole (Electrical) Inspection Checklist

VAULT AND MANHOLE (ELECTRICAL) INSPECTION CHECKLIST

Maintenance Inspector's Name ______________________________

Description and Location of Items Being Inspected ______________________________

____

Instructions: Perform the specified inspection and maintenance tasks. Insert the date when the activity is complete. Make any comments which are pertinent to future maintenance needs.

Item to Be Inspected	Date	Comments
Check manhole covers and gratings for plugged vent, defective gaskets, cracks, rust, corrosion, fit, and other structural deficiencies.		
Check vault doors for freedom of operation; broken hinges, locks, or latches; rust, corrosion, abrasions, or other deficiencies.		
Check ladders for missing or broken rungs or members; rot, rust, corrosion, or other deficiencies or unsafe condition.		
Check for dangerous, obnoxious, or flammable gases.		
Check lights and switches for broken or missing parts and proper operation.		
Check operation of pump.		
Check ground wire connection.		
Inspect cables for cracks, punctures, and other damage.		
Check fireproofing cable insulation for security.		
Check potheads and junction boxes for rust, corrosion, and deterioration.		

Provide any suggestions or recommendations for follow-up activities or any other comments related to the inspection and maintenance activities.

7.17: Switch Gear Inspection Checklist

SWITCH GEAR INSPECTION CHECKLIST

Maintenance Inspector's Name ____________________

Description and Location of Items Being Inspected ____________________

____

Instructions: Perform the specified inspection and maintenance tasks. Insert the date when the activity is complete. Make any comments which are pertinent to future maintenance needs.

Item to Be Inspected	Date	Comments
Inspect for evidence of overheating and looseness of connections on all mechanisms. Tighten as required.		
Inspect for rust and corrosion. Remove rust and corrosion and apply paint as applicable.		
Inspect for proper functioning of all moving parts. Lubricate as required.		

Provide any suggestions or recommendations for follow-up activities or any other comments related to the inspection and maintenance activities.

7.18: Telephone Line (Open Wire) Inspection Checklist

TELEPHONE LINE (0PEN WIRE) INSPECTION CHECKLIST

Maintenance Inspector's Name ______________________________

Description and Location of Items Being Inspected ______________________________

____

Instructions: Perform the specified inspection and maintenance tasks. Insert the date when the activity is complete. Make any comments which are pertinent to future maintenance needs.

Item to Be Inspected	Date	Comments
Check for unauthorized attachments on poles.		
Check clearances over private and public property; waterways, streets, driveways, alleys, and sidewalks.		
Check clearances from electric light and power lines, trees, trolley feeders, contact wires, or transformers including supporting structures.		
Check wires for proper sag and debris on wires.		
Check overall condition of telephone wire.		
Check open wire dead end for proper and secure construction.		
Check bridle cables and wires for loose connections, abraded insulation, kinks, uninsulated splices, proper placement, and termination.		
Check for proper connections, wire size in spans, and line wire joints.		
Check for strain on attachments, wires, and ties.		
Check for handmade or other unauthorized or obsolete splices, joints, or connections.		
Check for missing, broken, or incorrect insulators.		
Check condition at cable terminals, binding posts, bridging, and test connectors.		
Check condition of bridle wire insulators.		

Provide any suggestions or recommendations for follow-up activities or any other comments related to the inspection and maintenance activities.

7.19: Time Clock Inspection Checklist

TIME CLOCK INSPECTION CHECKLIST

Maintenance Inspector's Name ____________________

Description and Location of Items Being Inspected ____________________

____

Instructions: Perform the specified inspection and maintenance tasks. Insert the date when the activity is complete. Make any comments which are pertinent to future maintenance needs.

Item to Be Inspected	Date	Comments
Check security of installation. Tighten loose bolts or screws as required.		
Lubricate mechanism as required. *Do not over lubricate.*		
Check condition of ribbon and replace if requested.		
Check cleanliness of unit and clean as required.		
Inspect unit for excessive wear or damage.		
Inspect wiring for loose connections; charred, broken, or wet insulation; evidence of short-circuiting. Tighten, repair, or replace as required.		
Check clock for proper operation.		

Provide any suggestions or recommendations for follow-up activities or any other comments related to the inspection and maintenance activities.

7.20: Disconnecting Switch Inspection Checklist

DISCONNECTING SWITCH INSPECTION CHECKLIST

Maintenance Inspector's Name ______________________________

Date of Inspection ____________________ Location ____________________

Item Being Inspected ______________________________

OPERATING GEAR

____ Group-Operated Switches: rust, corrosion, loose brackets and holding bolts, nonrigid bearings and supports.

____ Grounding Cables, Clamps, and Straps: weak supports, broken or frayed portions of conductors, loose connections.

____ Insulating Section of Operating Rod: indications of cracks or signs of flashovers.

____ Movable Connections: inadequate lubrication, rust, corrosion, other conditions resulting in malfunctioning.

____ Switch: gears stiff or adjustment needed.

____ Locking and Interlocking Devices and Mechanisms: functional inadequacy to prevent unauthorized operation.

MOUNTINGS AND BASES

____ Rust, corrosion; twisted, bent, or warped; loose or missing ground wire.

INSULATORS

____ Cracks, Breaks, Chips, or Checking of Porcelain Glaze: more than thin or transparent film of dirt, dust, grease, or other deposits on porcelain.

____ Damage indicated by streaks of carbon deposits from flashovers.

____ Loose, broken, or deteriorated cement holding insulator to other parts.

BLADES AND CONTACTS

____ Excessive discoloration from overheating; roughness and pitting from arcing.

____ Misalignment of blades with contacts.

____ Arcing Horn Contacts: burns, pits, failure to contact each other throughout their length when switch is opened and closed.

____ Inadequate tension of bolts and springs.

____ Inadequate blade stop.

____ Lack of hinge lubrication; insufficient nonoxide grease for blades and contacts.

CONNECTIONS

____ Cable or Other Electrical Connections: loose bolts, discolorations indicating excessive heating at connection points.

____ Corrosion, particularly that resulting from atmospheric conditions.

7.20: *(Continued)*

CONNECTIONS (*continued*)

_____ Electrical Clearances of Cable or Other Conductor: inadequate to other phases or to ground for applicable circuit voltage.

_____ Flexible Connections: frayed, broken, or brittle.

_____ Cable from Grounding Switch to Grounding System: frayed, broken strands, loose connections.

COMMENTS:

7.21: Electronic Air Cleaner Inspection Checklist

ELECTRONIC AIR CLEANER INSPECTION CHECKLIST

Maintenance Inspector's Name ____________________

Description and Location of Items Being Inspected ____________________

____

Instructions: Perform the specified inspection and maintenance tasks. Insert the date when the activity is complete. Make any comments which are pertinent to future maintenance needs.

Item to Be Inspected	Date	Comments
Clean air filter and dry. Clean coils.		
Lubricate electric motor if necessary.		
Inspect wiring and electric controls for loose connections; tighten, repair, or replace if necessary.		
Inspect for rust and corrosion. Remove rust and corrosion and apply paint when applicable.		
Inspect guards and covers. Tighten or adjust, if necessary, to ensure secure fit.		
Replace charcoal filter as required.		

Provide any suggestions or recommendations for follow-up activities or any other comments related to the inspection and maintenance activities.

7.22: Battery Inspection Checklist

BATTERY INSPECTION CHECKLIST

Maintenance Inspector's Name ____

Description and Location of Items Being Inspected __________________________

____

Instructions: Perform the specified inspection and maintenance tasks. Insert the date when the activity is complete. Make any comments which are pertinent to future maintenance needs.

Item to Be Inspected	Date	Comments
Check battery for loose connections, corroded or dirty terminals. Clean and tighten connections as required.		
Check battery for proper water level and add distilled water as required.		
Check battery chargers and controls for proper operation.		
Inspect wiring and electrical controls for loose connections; charred, broken, or wet insulation; evidence of short-circuiting and other deficiencies. Tighten, repair, or replace as required.		
Check accuracy of instruments.		
Inspect battery room or enclosure for cleanliness, adequacy of ventilation; condition of floor; lighting and power fixtures to minimize fire hazard.		

Provide any suggestions or recommendations for follow-up activities or any other comments related to the inspection and maintenance activities.

7.23: Lightning Arrestor Inspection Checklist

LIGHTNING ARRESTOR INSPECTION CHECKLIST

Maintenance Inspector's Name ______________________________

Description and Location of Items Being Inspected ______________________________

____

Instructions: Perform the specified inspection and maintenance tasks. Insert the date when the activity is complete. Make any comments which are pertinent to future maintenance needs.

Item to Be Inspected	Date	Comments
Check line lead and ensure that it is securely fastened to line conductor and arrestor.		
Check to ensure that ground lead is securely fastened to arrestor terminal and ground.		
Ensure that arrestor housing is clean and free from cracks, chips, or evidence of flashover. Replace as necessary.		
Ensure that external gaps, if present, are free from foreign objects and are set at proper setting.		

Provide any suggestions or recommendations for follow-up activities or any other comments related to the inspection and maintenance activities.

7.24: Emergency Generator Set (Diesel or Gasoline) Inspection Checklist

**EMERGENCY GENERATOR SET (DIESEL OR GASOLINE)
INSPECTION CHECKLIST**

Maintenance Inspector's Name ________________________________

Description and Location of Items Being Inspected ________________________________

____

Instructions: Perform the specified inspection and maintenance tasks. Insert the date when the activity is complete. Make any comments which are pertinent to future maintenance needs.

Item to Be Inspected	Date	Comments
Check engine oil, fuel, radiator water, and battery water, and add or change if necessary.		
Start and run unit for 30 minutes; cycle generator through load range; observe mechanical and electrical operation and make minor adjustments as required.		
Clean unit and filters as required.		
Inspect electrical wiring, connections, switches, brushes, contacts, etc., and repair, replace, tighten, or adjust as required.		
Lubricate as required.		
Inspect for rust and corrosion. Remove rust and corrosion and apply paint where applicable.		

Provide any suggestions or recommendations for follow-up activities or any other comments related to the inspection and maintenance activities.

7.25: Emergency Lighting Inspection Checklist

EMERGENCY LIGHTING INSPECTION CHECKLIST

Maintenance Inspector's Name ______________________________

Description and Location of Items Being Inspected ______________________________

____

Instructions: Perform the specified inspection and maintenance tasks. Insert the date when the activity is complete. Make any comments which are pertinent to future maintenance needs.

Item to Be Inspected	Date	Comments
Safety—comply with all current precautions.		
Check for loose connections and broken wires. Tighten loose connections and repair or replace broken wires.		
Test pilot and power lights.		
Test transformer and rectifier (voltage).		
Check battery water level and add water necessary to keep water level above plates.		
Clean oxidation from terminals and test for broken and loose connections.		

Provide any suggestions or recommendations for follow-up activities or any other comments related to the inspection and maintenance activities.

7.26: Doors (Power-Operated) Inspection Checklist

DOORS (POWER-OPERATED) INSPECTION CHECKLIST

Maintenance Inspector's Name ______________________________

Description and Location of Items Being Inspected ______________________________

____

Instructions: Perform the specified inspection and maintenance tasks. Insert the date when the activity is complete. Make any comments which are pertinent to future maintenance needs.

Item to Be Inspected	Date	Comments
Check chain drive or belts for excessive slack and adjust or replace as applicable.		
Lubricate electric motor and chain drive as applicable.		
Inspect wiring and electrical controls for loose connections; charring, broken, or wet insulation; evidence of short-circuiting and other deficiencies. Tighten, repair, or replace as required.		
Lubricate and adjust door tracks/bearings as required.		
Inspect for rust and corrosion. Remove rust and corrosion and apply paint where applicable.		
Check operation of door safety devices, limit switches, warning alarm, etc., as applicable.		

Provide any suggestions or recommendations for follow-up activities or any other comments related to the inspection and maintenance activities.

7.27: Electrical Pothead Inspection Checklist

ELECTRICAL POTHEAD INSPECTION CHECKLIST

Maintenance Inspector's Name ______________________________

Description and Location of Items Being Inspected ______________________________

____

Instructions: Perform the specified inspection and maintenance tasks. Insert the date when the activity is complete. Make any comments which are pertinent to future maintenance needs.

Item to Be Inspected	Date	Comments
Check porcelain for cracks, breaks, chips, checking of porcelain glaze, carbon deposits, dirt, dust, grease, deterioration of cement sealing compound, and leakage or signs of moisture.		
Check cable clamps for corrosion, loose connections, and bolts.		
Check terminal studs and bolting pads for corrosion, loose connections, or poor contacts.		
Check mountings for corrosion and other defects.		

Provide any suggestions or recommendations for follow-up activities or any other comments related to the inspection and maintenance activities.

7.28: Electrical Power Plant Inspection Checklist

ELECTRICAL POWER PLANT INSPECTION CHECKLIST

Maintenance Inspector's Name ____________________________

Description and Location of Items Being Inspected ________________________

____

Instructions: Perform the specified inspection and maintenance tasks. Insert the date when the activity is complete. Make any comments which are pertinent to future maintenance needs.

Item to Be Inspected	Date	Comments
Check power plant for cleanliness and orderliness.		
Check for proper posting of safety signs and operating instructions.		
Check for operating log, plant log, and maintenance records.		
Test generator field, armature winding, and cable insulation.		
Check adequacy, serviceability, and reliability of generator excitation system.		
Check condition of emergency exciters and associated equipment including rheostats, pilot exciters, voltage regulators, and motor drives.		
Check ground-indicating system in ungrounded exciter circuits.		
Inspect wiring and electrical controls for loose connections; charred, broken, or wet insulation; evidence of short-circuiting and other deficiencies. Tighten, repair, replace as required.		

Provide any suggestions or recommendations for follow-up activities or any other comments related to the inspection and maintenance activities.

7.29: Electrical System (Waterfront) Inspection Checklist

ELECTRICAL SYSTEM (WATERFRONT) INSPECTION CHECKLIST

Maintenance Inspector's Name ______________________________

Description and Location of Items Being Inspected ______________________________

____

Instructions: Perform the specified inspection and maintenance tasks. Insert the date when the activity is complete. Make any comments which are pertinent to future maintenance needs.

Item to Be Inspected	Date	Comments
Check conductor enclosures and supports for corrosion, damage, missing or loose covers and fittings, proper drainage, dirt and debris, and missing fasteners.		
Check conductors for orderly arrangement, adequate support, proper sag, and adequate insulation.		
Check insulation for damage, signs of arcing, and hot spots.		
Check outlets for condition, signs of arcing, loose or missing fittings and covers, and corrosion.		
Check switches and breakers for condition, alignment of contacts, and signs of arcing.		
Inspect wiring and electrical controls for loose connections; charred, broken, or wet insulation; evidence of short-circuiting and other deficiencies. Tighten, repair, or replace as required.		

Provide any suggestions or recommendations for follow-up activities or any other comments related to the inspection and maintenance activities.

7.30: Oil Circuit Breaker Inspection Checklist

OIL CIRCUIT BREAKER INSPECTION CHECKLIST

Maintenance Inspector's Name ____________________

Description and Location of Items Being Inspected ____________________

____

Instructions: Perform the specified inspection and maintenance tasks. Insert the date when the activity is complete. Make any comments which are pertinent to future maintenance needs.

Item to Be Inspected	Date	Comments
Check all connections and tighten or adjust as required.		
Check resistance characteristics of the circuit-breaker contacts with ducter and record readings.		
Check insulating oil for contamination and resistivity. Filter or refill with new oil as required.		
Check oil level and add oil as necessary to obtain proper level.		
Check oil valves and gaskets for leakage. Repair and/or replace as required.		

Provide any suggestions or recommendations for follow-up activities or any other comments related to the inspection and maintenance activities.

7.31: Navigation Light Inspection Checklist

NAVIGATION LIGHT INSPECTION CHECKLIST

Maintenance Inspector's Name ______________________________

Description and Location of Items Being Inspected ______________________________

____

Instructions: Perform the specified inspection and maintenance tasks. Insert the date when the activity is complete. Make any comments which are pertinent to future maintenance needs.

Item to Be Inspected	Date	Comments
Check light for proper operation.		
Remove dirt accumulation as required.		
Check for rust and corrosion. Remove rust and corrosion and apply paint where applicable.		
Inspect wiring and electrical controls for loose connections; charred, broken, or wet insulation; evidence of short-circuiting and other deficiencies. Tighten, repair, or replace as required.		

Provide any suggestions or recommendations for follow-up activities or any other comments related to the inspection and maintenance activities.

7.32: Fire Alarm Panel Inspection Checklist

FIRE ALARM PANEL INSPECTION CHECKLIST

Maintenance Inspector's Name ________________________________

Description and Location of Items Being Inspected ________________________________

____

Instructions: Perform the specified inspection and maintenance tasks. Insert the date when the activity is complete. Make any comments which are pertinent to future maintenance needs.

Item to Be Inspected	Date	Comments
Check system for proper operation.		
Replace burned-out light bulbs/fuses as necessary.		
Check batteries and burglar alarm as required.		
Inspect for loose or inadequate connections; tighten, repair, or replace as necessary.		
Remove rust and corrosion and apply paint as applicable.		

Provide any suggestions or recommendations for follow-up activities or any other comments related to the inspection and maintenance activities.

7.33: Fire Alarm Box Light Inspection Checklist

FIRE ALARM BOX LIGHT INSPECTION CHECKLIST

Maintenance Inspector's Name ______________________________

Description and Location of Items Being Inspected ______________________________

____

Instructions: Perform the specified inspection and maintenance tasks. Insert the date when the activity is complete. Make any comments which are pertinent to future maintenance needs.

Item to Be Inspected	Date	Comments
Check for cracked or broken luminaries and fixture parts, missing pull-cords, and insulating links on pull-chains.		
Inspect wiring and sockets for loose connections, broken insulation, and other damage.		
Check fire alarm box lights; location and adequate support.		

Provide any suggestions or recommendations for follow-up activities or any other comments related to the inspection and maintenance activities.

7.34: Fire Alarm Box Inspection Checklist

FIRE ALARM BOX INSPECTION CHECKLIST

Maintenance Inspector's Name ______________________________

Designation and Location of Item Inspected ______________________________

____

Instructions: Perform the specified inspection and maintenance tasks. Insert the date when the activity is complete. Make any comments which are pertinent to future maintenance needs.

Item to Be Inspected	Date	Comments
Check to see that all wiring is tight and in good condition.		
Check for water leaks.		
Oil mechanism in boxes if necessary.		
Check box ground to ensure that it is not tied to another ground system.		
Check box and mechanisms for possible lightning damage.		
Check terminal connections for possible corrosion. Remove corrosion.		
Check master box to see that it operates on all three folds. A. Alarm records on metallic system.		
B. Alarm records on negative wire.		
C. Alarm records on positive wire.		
D. Alarm records on ground.		
Check speed of box.		
Check fire box light to ensure proper operation.		
REMOTE FIRE ALARM BOXES Check to see that all wiring is tight and in good condition.		
Check remote box for water leaks.		
Check switches for adjustment. Adjust as required.		
Actuate box and check to see that the remote will actuate the master box.		

Provide any suggestions or recommendations for follow-up activities or any other comments related to the inspection and maintenance activities.

7.35: Food Warmer/Grill Inspection Checklist

FOOD WARMER/GRILL INSPECTION CHECKLIST

Maintenance Inspector's Name ______________________________

Designation and Location of Item Inspected ______________________________

____

Instructions: Perform the specified inspection and maintenance tasks. Insert the date when the activity is complete. Make any comments which are pertinent to future maintenance needs.

Item to Be Inspected	Date	Comments
Inspect wiring and electrical controls for loose connections; charred, broken, frayed, or wet insulation; evidence of short-circuiting, defective operation and other deficiencies. Tighten, repair, and replace as required.		
Lubricate motor(s) as required. *Do not over lubricate.*		
Check motor(s) for excessive heat and vibration.		
Inspect piping, coils, etc., for loose connections, leaks, and corrosion. Tighten, repair, or replace as required.		
Inspect thermal insulation and protective coverings for open seams, missing sections, and loose fastenings. Tighten, repair, and replace as required.		
Check covers, lids, or doors for proper fit, and condition of gaskets/seals. Replace or repair as required.		
Check operating controls for proper operation through complete cycle. Repair or adjust as required.		
Inspect for rust and corrosion. Remove rust and corrosion and apply paint where applicable.		

Provide any suggestions or recommendations for follow-up activities or any other comments related to the inspection and maintenance activities.

7.36: Fuser Disconnect Inspection Checklist

FUSER DISCONNECT INSPECTION CHECKLIST

Maintenance Inspector's Name ____________________

Designation and Location of Item Inspected ____________________

____

Instructions: Perform the specified inspection and maintenance tasks. Insert the date when the activity is complete. Make any comments which are pertinent to future maintenance needs.

Item to Be Inspected	Date	Comments
Check fuse tubes and replace any that are severely eroded or burned on the inside or that are warped, broken, scorched, or burned.		
Check contacts and replace if annealed due to excessive heat or if badly pitted.		
Clean accumulated dirt and contamination from tubes.		
Smooth rough or slightly pitted contacts using a file or fine sandpaper.		
Clean corrosion and dirt from contacts and coat with nonoxidizing grease or paste.		
Check fuse tube and clip to ensure proper alignment and pressure.		

Provide any suggestions or recommendations for follow-up activities or any other comments related to the inspection and maintenance activities.

7.37: Underground Cable Inspection Checklist

UNDERGROUND CABLE INSPECTION CHECKLIST

Maintenance Inspector's Name ____________________

Designation and Location of Item Inspected ____________________

____

Instructions: Perform the specified inspection and maintenance tasks. Insert the date when the activity is complete. Make any comments which are pertinent to future maintenance needs.

Item to Be Inspected	Date	Comments
Check manholes for loose fit or missing covers; evidence of flooding or excessive moisture; water seepage through walls, floors, and around duct entrances.		
Check underground cable trench for pronounced depressions indicating a drop in the trench base.		
Check cable racks and ties for looseness and corrosion.		
Check cables for corrosion.		

Provide any suggestions or recommendations for follow-up activities or any other comments related to the inspection and maintenance activities.

C. SPECIAL FORMS

There are many maintenance-related activities in every company that must be included in the organization's overall management program. Most of these tasks are unique to the company. To assist management, special forms can be developed for the specific activity to enable them to control the respective maintenance function more effectively and efficiently.

Every company should evaluate its own operations to ascertain where such forms would help to make the management function more efficient. Once these activities are identified, appropriate documents can be developed. Some examples relating to housekeeping activities are included in this part of the section.

The Notice of Electrical Power Shutdown is used to communicate to personnel when power won't be available so they can make alternate plans. The Motor Job sheet is used to document the results of testing a motor. In order to implement an efficient light bulb or tube replacement program, it is necessary to keep track of how often they need to be and/or are changed. The Light Tube Replacement and Light Service Requisition forms can be used for this purpose. The last special form included in this section is the Elevator Call Checklist. It is used in the activity of servicing or repairing an elevator.

All of these forms are completed by providing the information requested and inserting it in the spaces provided. They are then transmitted to the designated management personnel and/or filed for future use in the control of the maintenance function.

7.38: Notice of Electrical Power Shutdown

NOTICE OF ELECTRICAL POWER SHUTDOWN

Power circuit to be de-energized		
Date of outage		
Estimated time of outage:	Off:	
	On:	
Circuits to be switched		
Areas affected		
Reason		
Requested by		
Date		
Switching to be supervised by		
Special Precautions		
Approved (plant engineer)		Date

FORM 7.38

Key Use of Form:

To notify affected personnel that electrical power will be disconnected.

Who Prepares:

Supervisor disconnecting power to perform maintenance activities.

Who Uses:

Department personnel being affected by power outage.

How to Complete:

Provide the information requested in the appropriate spaces.

Alternative Forms:

A company can revise this form to meet its specific needs.

7.39: Motor Job Sheet

MOTOR JOB SHEET

Motor No. ____________ Date ____________

Motor Home ________ Manufacturer ________ Model ________

Style ______ Frame ______ Phase ______ Volts ______ Amps ______

Cyc. ______ HP ______ Ser. No. ______ Form ______ RPM ______

°Cent. Rise ______ Rotor or Secondary Volts ______ Sec. Amps ______

Other ____

____

Megger Test	**Motor Location after Repairs or Movement**
Field—Stator ____________	____________
Armature—Rotor ____________	____________

Work Performed—Check Items

Job Order No. ______ New Fields ______ Relead ______ Rewind ______

New Shaft ____________ New Bearings ____________

Shaft End No. ____________ Stock? ____________

Comm. End No. ____________ Stock? ____________

Commutator Repair ______ Turn and Undercut ______ New Brushes ______

Windings Painted ______ Turn Slip Rings ______ Gear Case Repair ______

Pulley or Gear with Motor ______ Size ______ Total Labor Hours ______

Comments: ____

____

____

Supervisor ____________________

By ____________________

7.40: Light Tube Replacement Form

LIGHT TUBE REPLACEMENT FORM

Supervisor ______________________

Date	Floor	Bldg.	Total	Date	Floor	Bldg.	Total	Date	Floor	Bldg.	Total	Date	Floor	Bldg.	Total	Date	Floor	Bldg.	Total

FORM 7.40

Key Use of Form:

To document the replacement of light tubes.

Who Prepares:

Supervisor responsible for performing tube replacement.

Who Uses:

Designated personnel in maintenance department or plant engineering to evaluate efficiency of tubes and to plan for any needed changes in the tube type, manufacturer, or time between changes.

How to Complete:

Provide the information requested in the spaces provided.

Alternative Forms:

None.

7.41: Light Service Requisition

LIGHT SERVICE REQUISITION

Supervisor ________________________

Building	Floor	Type	Date	Location	Comments

FORM 7.41

Key Use of Form:

To order work on the lighting system, specifically the replacement of burners or bulbs or tubes.

Who Prepares:

Department supervisor in area where lights need to be replaced.

Who Uses:

Appropriate maintenance department personnel who writes work order to have lights replaced.

How to Complete:

Provide the information requested in the appropriate spaces.

Alternative Forms:

Form can be revised to make company-specific.

7.42: Elevator Call Checklist

ELEVATOR CALL CHECKLIST

1. Date ________________
2. Elevator Designation ________________
3. Stopped on which level ________________
4. Doors (as found) ______ open
   ______ closed
   ______ off track
   ______ other ________________
5. Stop button (red, top)—(as found) ______ in
   ______ out
   ______ unknown due to closed doors
6. Time you were notified ________ hours
7. By whom ________________
8. Time elevator contractor called ________ hours
9. Spoke to ______ answering service
   ______ mechanic
   ______ other ________________
10. Downtime was most likely caused by ______ personnel or equipment
    ______ elevator equipment
    ______ unknown

COMMENTS:

By: ________________

FORM 7.42

Key Use of Form:

To document a problem with an elevator and communication with the elevator repair contractor.

Who Prepares:

Maintenance personnel receiving the complaint and making the call to the contractor.

Who Uses:

Maintenance management to keep track of problems with elevators and use in future decision making related to maintenance, repair, and possible replacement of elevators.

How to Complete:

Provide the information requested in the appropriate spaces.

Alternative Forms:

Form can be revised to meet specific company needs.

Section 8

MAINTENANCE MANAGEMENT

Without an effective maintenance organization, the maintenance budget will get out of control, leading to premature breakdown and failure of buildings, systems, and/or equipment. This section will provide the manager with forms that can be used to keep the maintenance function under control and thus operating efficiently.

A. INVENTORY FORMS

The Maintenance Specification and Inventory Record forms given are used for almost any kind of equipment found in a building. To keep track of maintenance activities performed on building components and/or equipment, the Equipment Maintenance Record and Equipment History forms can be used.

8.1: Maintenance Specification

MAINTENANCE SPECIFICATION

Name ______________________________ Company No. ____________________

Location—Bldg. _____ Floor _____ Dept. _____ Cost Classification ____________________

Dwg. or Serial No. ____________________ Mfgr. ____________________

______________________________ Mfgr. No. ____________________

Type ____________ Size ____________ Vendor ____________________

HP ________ Volt ________ Amps ________ P.O. = ________ N.W. = ________

Phase ________ Cycle ________ RPM ________ Original Cost ____________________

Frame ________ Ratio ________ Cap. ________ Eng. Service ____________________

Class ________ Series ________ Weight ________ Foundation ____________________

Code ________ Style ________ Model ________ Installation ____________________

Auxiliary Equipment No.'s ____________________ Misc. ____________________

______________________________ Total Orig. Cost ____________________

Remarks ______________________________ Date Installed ____________________

______________________________ Addl. Charge ____________________

______________________________ Total ____________________

____

____

____

Inspection—Item ______________________________ Type ____________ Frequency ____________

____

____

____

____

Periodic Overhaul or Replacement—Item ______________________________

Description ____

______________________________ Frequency ______________________________

Lubrication—Item ______________________________ Type—Grade ____________ Frequency ____________

____

____

____

____

8.1: *(Continued)*

FORM 8.1

Key Use of Form:

To inventory equipment.

Who Prepares:

Designated maintenance personnel or plant engineering personnel.

Who Uses:

Plant engineering and maintenance department personnel.

How to Complete:

The information to be placed on this form is obtained from the specifications and drawings for the specific piece of equipment. For example, wiring documents are available from the manufacturer. Once completed the document is filed for future use when problems arise with the equipment. Furthermore, as inspections and other maintenance activities are performed, the results will be recorded on the form. This will provide the maintenance staff with an up-to-date record on the piece of equipment.

Alternative Forms:

Refer to similar inventory forms contained in Sections 6 and 7.

8.2: Inventory Record

INVENTORY RECORD

Division Name		Bldg. Name		Bldg. No.	
Equipment Description		Mfgr. Name			
Model No.	Serial No.		Req. No.	Fund No.	
Budget No.	Date	Fiscal Year	P.O. No.	Date Received	
Original Cost	Service Contract	Yes	No	Guarantee Expires	
Add to Inventory		Date	Delete from Inventory		Date
Installed in Building		Date	Location: Bldg. No.	Room No.	
REMARKS:					
		Submitted By			

FORM 8.2

Key Use of Form:

To inventory equipment.

Who Prepares:

Designated maintenance personnel or plant engineering personnel.

Who Uses:

Plant engineering and maintenance department personnel.

How to Complete:

The information to be placed on this form is obtained from the specifications and drawings for the specific piece of equipment. For example, wiring documents are available from the manufacturer. Once completed the document is filed for future use when problems arise with the equipment. Furthermore, as inspections and other maintenance activities are performed, the results will be recorded on the form. This will provide the maintenance staff with an up-to-date record on the piece of equipment.

Alternative Forms:

Refer to similar inventory forms contained in Sections 6 and 7.

8.3: Equipment Maintenance Record

EQUIPMENT MAINTENANCE RECORD

Equipment Designation		Location
Date Installed		Manufacturer
Manufacturer's Data		
Maintenance Performed	**Date**	**Person Performing Maint.**

FORM 8.3

Key Use of Form:

Record all maintenance activities.

Who Prepares:

Designated maintenance personnel.

Who Uses:

The form is retained by the central maintenance office for future use in controlling the maintenance for the specific piece of equipment.

How to Complete:

There will be one form for each piece of equipment. After a maintenance activity such as lubricating a fitting or inspecting the equipment has been performed, it will be so noted on this form. The actual completion of the maintenance activity will be known when the work order forms are returned from the supervisor assigned to perform the task(s).

Alternative Forms:

Refer to other inventory-type forms contained in Sections 6 and 7.

8.4: Equipment History

EQUIPMENT HISTORY

Equipment No. ________________ Description ________________________________

HISTORY OF ALTERATIONS AND REPAIRS					
			Cost		
Date	Job No.	Description of Work	Labor	Material	Total

FORM 8.4

Key Use of Form:

Record all maintenance activities. This form is similar to the previous one but is more detailed in that it has space in which the actual cost of performing each maintenance activity is recorded. This allows the company to monitor the cost of maintaining each piece of equipment. The cost data will be utilized in making productivity analysis of operations involving the specific equipment.

Who Prepares:

Designated maintenance personnel.

Who Uses:

The form is retained by the central maintenance office for future use in controlling the maintenance for the specific piece of equipment.

How to Complete:

There will be one form for each piece of equipment. After a maintenance activity such as lubricating a fitting or inspecting the equipment has been performed, it will be so noted on this form. The actual completion of the maintenance activity will be known when the work order forms are returned from the supervisor assigned to perform the task(s).

Alternative Forms:

Refer to other inventory-type forms contained in Sections 6 and 7.

B. MAINTENANCE WORK REQUEST FORMS

To initiate maintenance activities such as inspections, service, and repairs, work orders or requests are prepared. This section includes a series of different types of work order forms.

The first form, Immediate Maintenance Request, is to be used to request emergency maintenance. This is usually required in cases where one's safety or health is jeopardized or company operations will come to a standstill if the maintenance isn't performed. The next three forms, Repair or Service Order, Maintenance Service Request, and Minor Maintenance Work Order are basic work requests used to initiate service-type or minor maintenance operations such as lubricating a motor or fixing a door. These are simple to complete and provide very little information in terms of resources needed to complete the job.

If management needs specific information on what resources were needed and/or were used and the cost of performing the maintenance tasks, more detailed work order forms would be used. The ones included in this section are Maintenance Work Order, Maintenance Work Request, and the Work Order.

8.5: Immediate Maintenance Request

IMMEDIATE MAINTENANCE REQUEST

Charge Acct. No. ______________________ Department ______________________

To (Department) ______________________________________ Date ________________

Equipment: Number ________________________ Down ________________________

Location: Building ______________ Floor ______________ Bay ______________

Problem ____

____

Requested By ____________ Phone Ext. ____________ Accepted By ____________

Service Dept.	Clock No.	Time In	Time Out	Work Hours

Date Completed ______________ Total ____________

Explain what was done ____

____

____

Parts replaced ____

____

Approved by ____

____

FORM 8.5

Key Use of Form:

Requesting emergency maintenance.

Who Prepares:

Supervisor responsible for the operation of the equipment or item needing maintenance or person who is responsible for entering work requests.

Who Uses:

The central maintenance organizational unit uses this form for the purpose of noting the completion of the activity and filing it for possible future use in controlling the specific maintenance activity.

How to Complete:

The top part of the form is completed by the originator. It is then transmitted to the supervisor responsible for performing the maintenance. He or she will complete the balance of the form upon completion of the requested work.

Alternative Forms:

Specific revisions can be made to make the form more company-specific.

8.6: Repair or Service Order

REPAIR OR SERVICE ORDER

Dept. ______________________ Date ______________

Account Number ______________ Time ______________

Requested By ______________________________

Location Building ______________________________

Room No. ______________________________

Service or Repair Requested ______________________________

____

____

____

____

Work Completed By ______________________________

Date ______________________ Time ______________

Job. No. ______________________ Labor $ ______________

Materials ______________

Other ______________

TOTAL $ ______________

FORM 8.6

Key Use of Form:

To request a repair or service order.

Who Prepares:

Supervisor responsible for the operation of the equipment or item needing maintenance or person who is responsible for entering work requests.

Who Uses:

The central maintenance organizational unit uses this form for the purpose of noting the completion of the activity and filing it for possible future use in controlling the specific maintenance activity.

How to Complete:

The top part of the form is completed by the originator. It is then transmitted to the supervisor responsible for performing the maintenance. He or she will complete the balance of the form upon completion of the requested work.

Alternative Forms:

Specific revisions can be made to make the form more company-specific.

8.7: Maintenance Service Request

MAINTENANCE SERVICE REQUEST

Department	Date
Title	
Work Description	
Originator:	Department Head:
Department Priority:	Requested Completion Date:

FORM 8.7

Key Use of Form:

To order maintenance service.

Who Prepares:

Person making request.

Who Uses:

Central maintenance department to plan and schedule future maintenance activities.

How to Complete:

Provide the information requested on the form in the appropriate spaces.

Alternative Forms:

Form 8.5 or 8.6, or make specific revisions to it to make it meet specific company needs.

8.8: Minor Maintenance Work Order

MINOR MAINTENANCE WORK ORDER

Number:	Date Written:
Equipment Number:	Date Wanted:
Description of Job	
Signed:	

FORM 8.8

Key Use of Form:

To order minor maintenance work.

Who Prepares:

Person responsible for making request for work.

Who Uses:

Central maintenance department to plan and schedule future maintenance activities.

How to Complete:

Provide the information requested on the form in the appropriate spaces.

Alternative Forms:

Forms 8.5 through 8.7, or specific revisions to this form to make it more company-specific.

8.9: Maintenance Work Order

MAINTENANCE WORK ORDER

Date ______________________ Time ______________________

Area	Department	Bldg.	Floor	Equipment	Acct. No.

Priority ______________________

Emergency ______________ Urgent ______________ Can start later ______________

Description of work ____

____

____

____

Route to ____

___ Check prod. supv. before starting ___ Switch lockout ___ Protective Equip. Req.

Requested by ______________________ Approved by ______________________

Craft Sequence	Job Steps	No. Workers × Elapsed Hours	
		Estimated	Actual

Material and/or Equipment Requirements

Quan.	Cost	Description	Available	Ordered

Engineered by: ____________ Completed by: ____________ Accepted by: ____________

FORM 8.9

Key Use of Form:

To order maintenance work to be performed.

Who Prepares:

Person responsible for writing work orders along with those who have to approve the work.

Who Uses:

Supervisor responsible for doing the work and the responsible person in central maintenance or plant engineering to plan and control future maintenance activities relative to the specific item being maintained.

How to Complete:

Provide the information requested in the appropriate spaces.

Alternative Forms:

This form can be revised to make it company-specific. If simpler formats are needed, refer to Forms 8.5 through 8.8.

8.10: Maintenance Work Request

MAINTENANCE WORK REQUEST

Equipment ________________ Charge No. ________________ Date ________________

Originator ____ Approved ________________

Work Description ____

Specific Problem ____

____

____

Priority: 1 2 3

____

Work Plan: 1) ____

2) ____

3) ____

4) ____

5) ____

	Material Required	Qty.	Stk. No.	Reg. or P.O. No.	Del.
1)					
2)					
3)					
4)					
5)					
6)					

Labor:

Craft	No.	Hrs.	Shifts	Est. Total Work Hours

Work to Be Performed: During Operation ☐ Downshift ☐ W/E ☐

Priority: 1 2 3 Approved ________________

____

Record Following at End of Each Shift:

____

Problems: ____

____

____

8.10: *(Continued)*

Labor Record:

Date	Shift	Name	Start	Finish	Actual Hours
				TOTAL ACTUAL HOURS	

Issued to: 1st 2nd 3rd Date ______ Not Complete ☐ By ______ Complete ☐

FORM 8.10

Key Use of Form:

To order maintenance work to be performed.

Who Prepares:

Person responsible for writing work orders along with those who have to approve the work.

Who Uses:

Supervisor responsible for doing the work and the responsible person in central maintenance or plant engineering to plan and control future maintenance activities relative to the specific item being maintained.

How to Complete:

Provide the information requested in the appropriate spaces.

Alternative Forms:

Form 8.9, or add specific revisions to make it meet company needs.

8.11: Work Order

WORK ORDER

Account No. ______________________ Date ______________________

PRIORITY: Emergency ☐ Routine ☐ Standing ☐

ASSIGNED TO:

Carpentry	☐	Mech.–Plumbing	☐	Transportation	☐
Electric	☐	Painting	☐	Custodial	☐
Mech.–A/C–Refrig.	☐	Roof & Sheetmetal	☐	Security	☐
Mech.–Heat	☐	Grounds	☐	Other	☐

LOCATION: Building Name/Number ______________ Room _______ Area _______

ACCOMPLISH THE FOLLOWING WORK: ______________________________________

____

____

ESTIMATE FOR WORK ORDER ONLY:

LABOR

Mechanic(s) ______ hrs.

Helper(s) ______ hrs.

Materials List	Est. Cost	

FORM 8.11

Key Use of Form:

To order maintenance work to be performed.

Who Prepares:

Person responsible for writing work orders along with those who have to approve the work.

Who Uses:

Supervisor responsible for doing the work and the responsible person in central maintenance or plant engineering to plan and control future maintenance activities relative to the specific item being maintained.

How to Complete:

Provide the information requested in the appropriate spaces.

Alternative Forms:

Forms 8.9 and 8.10, or make appropriate revisions to the form to meet specific company needs.

C. WORK ORDER SUMMARY FORMS

To keep track of all the work orders issued, summary forms are presented on which each work order by date and/or number are recorded. This helps ensure a well-organized process of controlling the maintenance activities within a company.

The forms contained in this section are the Maintenance Work Sheet, Job Order Register, Labor and Material Work Order Summary, and Daily Service Log. Whatever format is used, as work orders are issued they should be recorded on the summary form. When the work has been completed, it should also be so noted along with any other information requested. The various forms differ in the amount of detail a company wishes to document for future use. Usually, the more detailed the work order, the more detailed the summary form should be.

The Labor and Material Work Order Summary, as the title indicates, is used to record the materials used and hours taken to perform the job. This information is used to determine the actual cost of performing the work and then comparing it to the estimated cost for the purpose of evaluating job efficiency.

8.12: Maintenance Work Sheet

MAINTENANCE WORK SHEET

Supervisor ______________________ Shift ______________ Date ______________

Acct. No.	Call Time	Status	Job Description	Supervisor Notified	Time Complete	Maint. Hours

8.13: Job Order Register

JOB ORDER REGISTER

Work Order No.	Date of W.O.	Dept.	Description	Assigned Supervisor	Date Comp.

8.14: Labor and Material Work Order Summary

LABOR AND MATERIAL WORK ORDER SUMMARY

Work Order No.	Description of Work	Type Work	Date Received	Material Cost	Date Completed	Estimated Hours	Actual Hours

8.15: Daily Service Log

DAILY SERVICE LOG

Log No.	Call Time	Location	Service Request	Hours		Crew	Time	
				Est.	Act.	Assigned	Start	Stop

D. INSPECTION FORMS

One of the purposes of a work order is to request an inspection or survey of some item or equipment. Several forms (Maintenance Survey, Plant Engineering and Maintenance Report, Maintenance Report, and Preventive Maintenance Inspection) have been included in this section on which the results of the survey or inspection can be made. These forms are completed by the inspector who provides the information requested in the appropriate spaces and forwards them to the designated maintenance personnel for recording on the inventory form and initiating follow-up action.

8.16: Maintenance Survey

MAINTENANCE SURVEY

Craft	Location	Inspector	Date

Item No.	Equipment or Facility	Action Required	Freq.

FORM 8.16

Key Use of Form:

To record results of inspection.

Who Prepares:

Person performing the inspection.

Who Uses:

Designated person in central maintenance management unit for use in planning and scheduling any needed follow-up work.

How to Complete:

Provide the information requested in the appropriate spaces.

Alternative Forms:

For inspecting specific items covered in the other sections of this book, use the appropriate form(s) provided. Forms 8.17–8.19 in this section may also apply. Furthermore, revisions could be made to this form to make it more company-specific.

8.17: Plant Engineering and Maintenance Survey Report

PLANT ENGINEERING AND MAINTENANCE SURVEY REPORT

Plant	Inspector
Date of Survey	
Type of Survey (Inspection)	
Type of Problem	
Location	
Problem	
Possible Cause(s) and Corrective Action	
PRIORITY	
Additional Comments	

FORM 8.17

Key Use of Form:

To record results of inspection.

Who Prepares:

Person performing the inspection.

Who Uses:

Designated person in central maintenance management unit for use in planning and scheduling any needed follow-up work.

How to Complete:

Provide the information requested in the appropriate spaces.

Alternative Forms:

For inspecting specific items covered in the other sections of this book, use the appropriate form(s) provided. Forms 8.16, 8.18, and 8.19 in this section may also apply. Furthermore, revisions could be made to this form to make it more company-specific.

8.18: Maintenance Report

MAINTENANCE REPORT

Maintenance Inspector's Name ______________________________

Date ________________ Crew ________________ Shift ________________

Inspections and Maintenance Checks and Adjustments	Inspected (Check ✓)	Lube OK	Lube Needed

REMARKS: (If additional space is needed, note here and use back of sheet.)

FORM 8.18

Key Use of Form:

To record results of inspection.

Who Prepares:

Person performing the inspection.

Who Uses:

Designated person in central maintenance management unit for use in planning and scheduling any needed follow-up work.

How to Complete:

Provide the information requested in the appropriate spaces.

Alternative Forms:

For inspecting specific items covered in the other sections of this book, use the appropriate form(s) provided. Forms 8.16, 8.17, and 8.19 in this section may also apply. Furthermore, revisions could be made to this form to make it more company-specific.

8.19: Preventive Maintenance Inspection

PREVENTIVE MAINTENANCE INSPECTION

Equipment ______________________ Location ______________________

Date Inspected ______________________ Inspector ______________________

Components or Parts to Be Inspected	Inspector's Initials	Comments

FORM 8.19

Key Use of Form:

Document results of preventive maintenance inspection.

Who Prepares:

Person ordering inspection and person performing the inspection.

Who Uses:

Designated personnel in central maintenance or plant engineering department to plan, schedule, and control future maintenance activities.

How to Complete:

This form is completed as a result of inspecting a specific piece of equipment. The person ordering the inspection will note what components or parts are to be inspected. This information is usually obtained from the manufacturer's maintenance recommendations. The next step is to transmit the form to the inspector. He or she will record their findings and return the document to the designated person who will record the completion of the activity on the specific inventory form and initiate any required follow-up tasks needed for the piece of equipment such as the replacement of a certain part.

Alternative Forms:

For inspecting specific items covered in the other sections of this book, use the appropriate form(s) provided. Forms 8.16–8.18 in this section may also apply. Furthermore, revisions could be made to this form to make it more company-specific.

E. INSPECTION SUMMARY FORMS

To keep track of the various inspections, summary forms are found next in this section. They are the Monthly Report of Preventive Maintenance Inspections and Inspection Record Sheet forms. Both of these are self-explanatory in their completion. They are generated in the central maintenance management office and retained there for future use in evaluating the performance of the maintenance organization.

8.20: Monthly Report of Preventive Maintenance Inspections

MONTHLY REPORT OF PREVENTIVE MAINTENANCE INSPECTIONS

Building Area	Inspection Scheduled	Inspection Completed	Inspection Incomplete	Jobs Resulting	Jobs Completed

8.21: Inspection Record Sheet

INSPECTION RECORD SHEET

Completed By ______________________ Inspection Period ______________________

Building	Item or System Inspected	Date	Remarks

F. PLANNING FORMS

Once work is ordered, the next thing is to plan, estimate, and schedule it. To assist the respective maintenance personnel in these tasks, appropriate forms have been included in this section. Some of these forms will be used together in cases where information from one form is needed on another, such as cost of materials on a planning form.

When planning a maintenance activity, one of three forms provided can be used. The three are the Maintenance Department Planner, Maintenance Planning Sheet, and Comprehensive Maintenance Planning Sheet. They vary in the amount of detail that is required not only in the planning process itself but also in the information needed to complete the respective form.

8.22: Maintenance Department Planner

MAINTENANCE DEPARTMENT PLANNER

Date ____________________

Please obtain quotations on:	
According to the attached specifications:	
We suggest:	
Quotations should be received by (date):	
Prepared by:	
Department:	Phone:
Remarks:	

FORM 8.22

Key Use of Form:

To obtain cost quotations to perform a specific maintenance activity.

Who Prepares:

Person doing the quote, such as a supervisor, planner, estimator, or other maintenance manager.

Who Uses:

Person in organization responsible for obtaining quotes. Once obtained, it is sent back for developing the estimate.

How to Complete:

Provide the information requested in the appropriate spaces.

Alternative Forms:

Company can make revisions to this form to meet its specific needs.

8.23: Maintenance Planning Sheet

MAINTENANCE PLANNING SHEET

Planner:	Department:	Estimated Hours By Skill	Date	Priority	Page ___ of ___
Contact Person:	Phone:				
Line:	Major Component:		Job No.	Work Order No.	Work Code / Charge No.
Job Description:					

Job Seq.	Description of Operation	No. Wk.	Skill	Duration Hours								Remarks

8.23: *(Continued)*

FORM 8.23

Key Use of Form:

Used to plan a job.

Who Prepares:

Planner or person responsible for planning.

Who Uses:

After the completion of the job, the information contained on the form is compared by the designated maintenance personnel against the actual resources it took to perform the activity to ascertain the efficiency of the operation. The form is then filed in a central location for possible future use.

How to Complete:

The planner provides the general kinds of information requested. Following this, he or she will divide the job into a series of activities and list them in sequential order in the space provided on the form along with the times, labor skills, and other resources needed to complete each step.

Alternative Forms:

Specific changes can be made to make the form more company-specific.

8.24: Comprehensive Maintenance Planning Sheet

COMPREHENSIVE MAINTENANCE PLANNING SHEET

Work Order No. ______________________________

Planner ______________________________ Date ______________

JOB: ______________________________

____

WORK SEQUENCE REQUIRED: 1st day: ______________

2nd day: ______________ 3rd day: ______________

4th day: ______________ 5th day: ______________

6th day: ______________ 7th day: ______________

(If more space is required, complete on back of this sheet.)

DRAWINGS, SKETCHES, OR REFERENCES: ______________

Craft/Schedule	Standard Time Allowances						
	1	2	3	4	5	6	7

(NOTE: For each day, indicate standard hours required to complete.)

SPECIAL TOOLS REQUIRED:

1) ______________ 4) ______________

2) ______________ 5) ______________

3) ______________ 6) ______________

SAFETY, EPA, OR OSHA REQUIREMENTS: ______________

____

____

WORKING CONDITIONS: ______________

____

MATERIALS REQUIRED:

From Storeroom: 1) ______________ 4) ______________

2) ______________ 5) ______________

3) ______________ 6) ______________

Purchased:

1) ______________ P.O.#: ______________ Delivery Date: ______________

2) ______________ P.O.#: ______________ Delivery Date: ______________

3) ______________ P.O.#: ______________ Delivery Date: ______________

4) ______________ P.O.#: ______________ Delivery Date: ______________

8.24: *(Continued)*

FORM 8.24

Key Use of Form:

Used to plan a job.

Who Prepares:

Planner or person responsible for planning.

Who Uses:

After the completion of the job, the information contained on the form is compared by the designated maintenance personnel against the actual resources it took to perform the activity to ascertain the efficiency of the operation. The form is then filed in a central location for possible future use.

How to Complete:

The planner provides the general kinds of information requested. Following this, he or she will divide the job into a series of activities and list them in sequential order in the space provided on the form along with the times, labor skills, and other resources needed to complete each step.

Alternative Forms:

Specific changes can be made to make the form more company-specific.

G. ESTIMATING FORMS

The next series of forms are used in the process of estimating and controlling the costs of performing the maintenance work. The Work Sheet is used to take off the type and quantity of the various materials needed. The Summary Sheet is where the estimator determines the cost of materials, labor, tools, and equipment to perform the work. In addition, a total cost is documented on this form. The Maintenance Cost Data Sheet is a simplified version of the Summary Sheet. Finally, the cost of performing various maintenance operations are summarized on the Recapitulation Sheet. After the work is complete it is important to determine if the actual cost of performing the work was equal to that of the estimate. The Estimate Breakdown form and Budget Control Sheet can be used in this cost analysis. It is also important to keep track of actual costs for maintaining such items as equipment. The Equipment Maintenance Cost Record has been designed for this purpose.

8.25: Work Sheet

WORK SHEET

By: | Date ________ Title: | Project ________ | Page No.
Sheet ____ of ____ ________________ W.O. No. ________

	Description	No.	Dimensions			Extensions	Subtotal			Total		Remarks
			Length	Width	Height		Qty.	Qt.		Qty.	Qt.	

8.25: *(Continued)*

FORM 8.25

Key Use of Form:

This form is used in the process of estimating the types and quantities of the various materials required to complete a specific maintenance operation.

Who Prepares:

Estimator.

Who Uses:

Estimating personnel, appropriate field supervisor(s), and people in central management for evaluation and control purposes.

How to Complete:

The estimator first provides the general information requested on the form. He then records the description of each material by kind and quality along with the overall dimensions (if appropriate). If any math operations are needed, such as multiplying two dimensions to obtain an area, they are shown under the extensions column. Finally, the estimator notes the sub and total quantities in the spaces provided.

Alternative Forms:

A company can change this form to meet its own specific needs. Also forms 8.27 and 8.29 could be used.

8.26: Summary Sheet

SUMMARY SHEET

By:	Date ______ Sheet ______ of ______					Title: ______				Project ______ W.O. No. ______			Page No.	
		Quantity		Material Cost		Labor Factors					Labor Cost		Item Cost	
							Work Hours							
	Description	Total	Qt.	Per Unit	Total	Craft	Per Unit	Total	Rate	Cost Per	Per	Total	Total	Per Unit
					Material							Labor	Total	

FORM 8.26

Key Use of Form:

This form is used to derive and record the cost of performing a specific task.

Who Prepares:

Estimator.

Who Uses:

Estimating personnel, appropriate field supervisor(s), and people in central management for evaluation and control purposes.

How to Complete:

The estimator notes the total material obtained from the work sheet. He or she then derives a total material cost by multiplying the total quantity by the unit material cost obtained from quotes of suppliers. The next step is to list the various crafts that will be used for the job along with their productivity and rate of pay. From this information the estimator will be able to calculate and record in the appropriate spaces the total time it will take to perform the job and the total labor cost. If equiment is required for the job, its cost should also be placed on this form. It can either be placed under the material or labor columns and a note made under the description column of the type and quantity of equipment being used. The form could also be revised to include an equiment column. The final step is to sum the total costs of material, labor, and equipment to obtain the total direct job cost. If a unit cost is desired, the estimator will divide the total cost by the total material quantity and place the number in the space provided.

Alternative Forms:

A company can change this form to meet its own specific needs. Also forms 8.27 and 8.29 could be used.

8.27: Maintenance Cost Data Sheet

MAINTENANCE COST DATA SHEET

Work Order No. ______________________ Job ______________________

Account No. ______________________ Prepared By ______________________

Labor				Material		
Emp. No.	**Hours**	**Date**	**Cost**	**Description**	**Date**	**Cost**
TOTAL LABOR COST $				TOTAL MATERIAL COST $		

Completed Date ______________________ Total Job Cost $ ______________________

8.27: *(Continued)*

FORM 8.27

Key Use of Form:

To obtain documented estimated job costs.

Who Prepares:

Estimator.

Who Uses:

Estimating personnel, appropriate field supervisor(s), and people in central management for evaluation and control purposes.

How to Complete:

This form is similar to Form 8.26 and should be completed in essentially the same manner but appropriate changes to the procedure to be consistent with the form should be made.

Alternative Forms:

The company can make appropriate changes to the form to meet its needs. Also, Forms 8.25 and 8.26 can be used in place of this one.

8.28: Recapitulation Sheet

RECAPITULATION SHEET

Date ______________________ Project ______________________

By ______________________ W.O. No. ______________________

Title ______________________ Sheet __________ of __________

Acct. No.	Item	Material	Labor	Sub.	Equipment	Total
	TOTALS					

FORM 8.28

Key Use of Form:

This document is used to summarize the costs of performing the various activities of a maintenance job.

Who Prepares:

Estimator.

Who Uses:

Estimating personnel, appropriate field supervisor(s), and people in central management for evaluation and control purposes.

How to Complete:

Using the information from the Summary Sheet (or similar form), the estimator notes the cost of the material, labor, equipment, and sub-contractor services (if not being performed by in-house personnel) for each activity. He or she then sums the total cost to perform each activity to derive a total direct job cost for the entire job.

Alternative Forms:

Company can make revisions to this form to meet its specific needs.

8.29: Estimate Breakdown

ESTIMATE BREAKDOWN

Project ______________________ Date ______________________

Prepared By ______________________

MATERIALS

Quantity	Description	Unit Cost	Totals

Sub Total ________

LABOR

Trade	Rate	Hours	Totals

Sub Total ________

OTHER

Contingencies ________

Grand Total ________

ANALYSIS OF ESTIMATE

Req. No. ________ Account No. ________ Est. Amt. ________ Date of Req. ________

Account No.	Date	Trade	Material	Labor	Total

Date work completed ______________ Total actual cost $ ______________

8.29: *(Continued)*

FORM 8.29

Key Use of Form:

To develop and document estimated costs and actual costs.

Who Prepares:

The estimate is prepared by the estimator and actual costs provided by the designated maintenance control personnel.

Who Uses:

Designated central maintenance personnel to evaluate efficiency of the work performed and the original estimate.

How to Complete:

Provide the information requested in the appropriate spaces both at the estimating stage and upon completion of the job.

Alternative Forms:

Specific revisions can be made to make the form more company-specific.

8.30: Budget Control Sheet

BUDGET CONTROL SHEET

Project Identification ______________________ Period Ending ______________

Project Number ______________ Reported By ______________________

Individual Work Items	Budget Hours A	% Total Budget $B = \frac{A}{\Sigma A}$	Hours Used C	% Budget Used $D = \frac{C(B)}{A}$	% Work Item Complete E	% Total Complete $B \times E$
PROJECT TOTALS						

FORM 8.30

Key Use of Form:

This document is used to analyze the use of hours budgeted during the estimating process. The reader should note that budgeted hours refers to estimated hours.

Who Prepares:

Designated person in plant engineering or the maintenance department.

Who Uses:

Designated maintenance management to evaluate the efficiency of the work being performed.

How to Complete:

In the maintenance organization, management must implement methods to constantly evaluate hours estimated to perform work against hours actually used. This form helps management do this. By following the instructions in each column, the person completing the form can determine the efficiency of the maintenance operation. If the job is not being performed to the standards established by the budget or estimate, the appropriate managerial personnel can take steps during the job to improve the productivity. Once completed, this form should also be filed for future use.

Alternative Forms:

Appropriate revisions can be made to make the form company-specific.

8.31: Equipment Maintenance Cost Record

EQUIPMENT MAINTENANCE COST RECORD

Machine No.	Type Machine	Catalog No.	Serial No.	
Vendor		Date Purchased	Date Installed	
Purchase Cost	Installation Cost	Total Cost	Depreciation	Year	Per Year

Depreciation Schedule					Mechanical Equipment			
Year	Depreciation	Added to Value	To Be Authorized	Maintenance Cost				
					Electrical Equipment			
					Equipment			
					Make			
					Serial No.			
					Type Frame			
					Voltage			
					Phase			
					Ampers			
					Horse Power			
					R.P.M.			
					Drive			
					Circuit			
					Date Installed			

8.31: *(Continued)*

FORM 8.31

Key Use of Form:

Record the actual cost of owning and operating equipment.

Who Prepares:

Personnel in the accounting department.

Who Uses:

Personnel in plant engineering department in performing an equipment replacement analysis.

How to Complete:

Provide the information requested in the appropriate space. The person completing the form can obtain much of the information from the manufacturer and in-house sources. Each year the cost of depreciation and maintenance is recorded on the form.

H. SCHEDULING FORMS

Once the work has been planned and estimated it needs to be scheduled, which is nothing more than placing dates on the various activities identified in the plan. To control the scheduling activities better, they need to be documented. There are various levels of scheduling: project, monthly, weekly, and daily. Each of these requires its own form.

The Project Master Schedule and Master Schedule forms have been provided to schedule an entire job prior to its start. The first form is less detailed than the second and should be used for smaller projects. Once a master schedule is prepared, the next step is to develop monthly, weekly, and daily schedules, where appropriate (that is, a monthly schedule would not be needed for a job that is estimated to take one week to complete). The Monthly Master Maintenance Log is to be used to schedule and log maintenance activities on equipment that utilizes oil in its operation. Similar forms can be developed for other types of items to be scheduled over a month's time. Also provided are the Weekly Maintenance Work Schedule, Daily Schedule, and Office Facilities Scheduled Maintenance Routine forms. As noted by their titles, they are used to schedule work over shorter durations. To keep track of scheduled work, the Schedule Control Sheet can be used to document the activities completed each week, which is a measure on how well the schedule was accomplished.

All of the schedule formats are essentially completed in the same manner in terms of the type of information that needs to be inserted on the form. All of them will be completed and administered by the designated person in the scheduling unit of the maintenance organization. Copies will most likely be provided to the supervisor(s) responsible for performing the work. The original will be retained by the scheduling office for use in evaluating the completion of the work and further scheduling.

8.32: Project Master Schedule

PROJECT MASTER SCHEDULE

Project Title ______________________ Date Prepared ______________

Prepared By ______________________ Functional Area ______________

Location ____

Task Description	Weeks Required	Calendar Dates	Remarks
1.			
2.			
3.			
4.			
5.			
6.			
7.			
8.			
9.			
10.			
11.			
12.			
13.			
14.			
15.			
16.			
17.			
18.			
19.			
20.			
21.			
22.			
23.			
24.			
TOTAL ESTIMATED TIME			

8.32: *(Continued)*

FORM 8.32

Key Use of Form:

Schedule maintenance activities.

Who Prepares:

Person responsible for preparing the schedule.

Who Uses:

Scheduling personnel and maintenance supervisors. Also, the appropriate maintenance manager will review the schedule to ascertain if the date and the actual time it took to perform a job was consistent with the schedule.

How to Complete:

Provide the information requested in the appropriate spaces.

Alternative Forms:

A company can revise the form to make it fit its specific needs. Also, Form 8.33 may be more applicable.

8.33: Master Schedule

MASTER SCHEDULE

Schedule Duration ______________________________

Date Completed ______________ Scheduled By ______________________

Work Order	Dept.	Operations	Worker	Est. Hours	Planned Start Date	Planned Date Compl.

FORM 8.33

Key Use of Form:

Schedule maintenance activities.

Who Prepares:

Person responsible for preparing the schedule.

Who Uses:

Scheduling personnel and maintenance supervisors. Also, the appropriate maintenance manager will review the schedule to ascertain if the date and the actual time it took to perform a job was consistent with the schedule.

How to Complete:

Provide the information requested in the appropriate spaces.

Alternative Forms:

A company can revise the form to make it fit its specific needs.

8.34: Monthly Master Maintenance Log

MONTHLY MASTER MAINTENANCE LOG

For Month of ____________________ Equipment Designation ____________________

Equipment Location ____________________ Form Completed By ____________________

Date	Hours Operated	Accumulated Hours Since Last Overhaul	Accumulated Hours Since Last Oil Change	Oil Consumption	Parts Used

FORM 8.34

Key Use of Form:

Schedule maintenance activities.

Who Prepares:

Person responsible for preparing the schedule.

Who Uses:

Scheduling personnel and maintenance supervisors. Also, the appropriate maintenance manager will review the schedule to ascertain if the date and the actual time it took to perform a job was consistent with the schedule.

How to Complete:

Provide the information requested in the appropriate spaces.

Alternative Forms:

A company can revise the form to make it fit its specific needs.

8.35: Weekly Maintenance Work Schedule

WEEKLY MAINTENANCE WORK SCHEDULE

Schedule Period Ending ______________________ Page _____ of _____

Work Listed by Priority by Department			Date, Day, and Shift							Promised Comp. Date	Remarks
Work Order Number	Work Code	Brief Description	_____	_____	_____	_____	_____	_____	_____		

FORM 8.35

Key Use of Form:

Schedule maintenance activities.

Who Prepares:

Person responsible for preparing the schedule.

Who Uses:

Scheduling personnel and maintenance supervisors. Also, the appropriate maintenance manager will review the schedule to ascertain if the date and the actual time it took to perform a job was consistent with the schedule.

How to Complete:

Provide the information requested in the appropriate spaces.

Alternative Forms:

A company can revise the form to make it fit its specific needs.

8.36: Daily Schedule

DAILY SCHEDULE

Date ______________________ Prepared By ______________________________________

Pertinent Information ____

			Work Order No.						
Name	Employee	Actual Time	Job Cost No.						
			TOTAL						

FORM 8.36

Key Use of Form:

Schedule maintenance activities.

Who Prepares:

Person responsible for preparing the schedule.

Who Uses:

Scheduling personnel and maintenance supervisors. Also, the appropriate maintenance manager will review the schedule to ascertain if the date and the actual time it took to perform a job was consistent with the schedule.

How to Complete:

Provide the information requested in the appropriate spaces.

Alternative Forms:

A company can revise the form to make it fit its specific needs.

8.37: Office Facilities Scheduled Maintenance Routine

OFFICE FACILITIES SCHEDULED MAINTENANCE ROUTINE

Date ____________ Account Number ____________

Title ____________ Page ________ of ________

Frequency	Number of Workers	Hours/Worker

Prepared by ____________

FORM 8.37

Key Use of Form:

Schedule maintenance activities.

Who Prepares:

Person responsible for preparing the schedule.

Who Uses:

Scheduling personnel and maintenance supervisors. Also, the appropriate maintenance manager will review the schedule to ascertain if the date and the actual time it took to perform a job was consistent with the schedule.

How to Complete:

Provide the information requested in the appropriate spaces.

Alternative Forms:

A company can revise the form to make it fit its specific needs.

8.38: Schedule Control Sheet

SCHEDULE CONTROL SHEET

Department ______________________ Name ______________________

Week Ending ______________________ Employee No. ______________________

Assignment Schedule, Check Off List and Performance Record

Location	Operation	Worker Time	Sched. Compl.

Mon.	Tues.	Wed.	Thurs.	Fri.

FORM 8.38

Key Use of Form:

Schedule maintenance activities.

Who Prepares:

Person responsible for preparing the schedule.

Who Uses:

Scheduling personnel and maintenance supervisors. Also, the appropriate maintenance manager will review the schedule to ascertain if the date and the actual time it took to perform a job was consistent with the schedule.

How to Complete:

Provide the information requested in the appropriate spaces.

Alternative Forms:

A company can revise the form to make it fit its specific needs.

I. PURCHASING, REQUISITION, AND EXPEDITING FORMS

Once the estimate to perform the specific maintenance function has been developed and the plan and schedule complete, the next step is to proceed to order the material. The material will either be purchased from sources outside the company or obtained in-house such as from storage. Exactly where it will be obtained from will depend on the magnitude and complexity of the job and the types of materials that are available in-house.

If the materials are obtained from external sources, a purchase order must be written. However, if the materials are coming from in-house sources, the document which is used to obtain them is called a requisition. A requisition is also used when in-house personnel, such as the supervisor of the maintenance department, needs some materials. He or she will write a requisition and send it to the purchasing department who in turn will either obtain the material from external sources with a purchase order or from within the company using a requisition form.

A function related to purchasing is known as expediting. This is the task of following up to ensure the correct material, in the quantity ordered, will arrive at the location specified in the purchase order or requisition on the specified date. This part of the section contains various types of forms used in performing the tasks described above, including that of inventorying (controlling) the materials, tools, and equipment once they are received.

8.39: Bill of Material

BILL OF MATERIAL

Work Description ______________________ Work Order No. ______________

Department ______________________ Project No. ______________

Line ______________________ Charge No. ______________

Major Component ____

Item No.	Quantity	Item Description	Status Code	Location	Price

STATUS CODES: S = Stock Item
SR = Stock Item—Replenish
OH = On Hand
O = Order for Project
OD = Already Ordered for Project

FORM 8.39

Key Use of Form:

This form is completed for the purpose of listing all the material required to complete the maintenance task and the cost of each different type.

Who Prepares:

Designated personnel in plant engineering or maintenance department that is responsible for developing estimate and/or ordering materials.

Who Uses:

Person ordering material.

How to Complete:

The person completing this form will use the Work Sheet, described earlier in this section, from which to acquire the needed quantities. The prices are obtained from the suppliers.

Alternative Forms:

A company can revise the form to make it fit its specific needs.

8.40: Maintenance Material Requisition

MAINTENANCE MATERIAL REQUISITION

From ______________________ P.O. No. ______________

______________________ Date ______________

____

Supplier ____

Address ____

Attention ______________________ Telephone ______________

Qt.	Unit	Description	Part No.	Amount

Requested by ______________________ Approved by ______________________

FORM 8.40

Key Use of Form:

This form is used to requisition materials from either internal or external sources.

Who Prepares:

Designated individual in plant engineering or maintenance department.

Who Uses:

Purchasing department to make material orders. If the material is to be obtained from in-house inventory, the form is sent to the appropriate organizational unit for fulfillment. Copies of the form are transmitted to the designated personnel and one kept in the permanent files.

How to Complete:

Provide the information requested in the appropriate spaces.

Alternative Forms:

Specific revisions can be made to make the form company-specific.

8.41: Material Order Form

MATERIAL ORDER FORM

Date ____________________

Job Description ______________________________ Log No. ____________

______________________________ Chg. No. ____________

Item No.	Quantity	Description of Material	Contract Number

FORM 8.41

Key Use of Form:

This form is used to requisition materials from either internal or external sources.

Who Prepares:

Designated individual in plant engineering or maintenance department.

Who Uses:

Purchasing department to make material orders. If the material is to be obtained from in-house inventory, the form is sent to the appropriate organizational unit for fulfillment. Copies of the form are transmitted to the designated personnel and one kept in the permanent files.

How to Complete:

Provide the information requested in the appropriate spaces.

Alternative Forms:

Specific revisions can be made to make the form company-specific.

8.42: Purchase Order

PURCHASE ORDER

From ______________________ Office ______________________

Address all invoices,
statements and correspondence
relative to this order to us at ______________________
(Name of Job)

Ordered from ______________________ ______________________

Address ______________________ ______________________

Req. No. P.O. No.

Deliver or ship to
Geo. E. Deatherage & Son at ______________________ ______________________

(This No. must appear on your invoice)

Care of ______________________ Ordered ____________ 19______

VIA ______________________ R.R. or Express Required ______________________

______________________ Switch ______________________

IMPORTANT: EACH ORDER MUST HAVE A SEPARATE INVOICE RENDERED IN TRIPLICATE. Invoices not in our office by the 3rd day of the month following date of delivery will not be paid until the following month. Terms: 2% deducted the 15th of the month following delivery unless otherwise agreed.

COST DISTRIBUTION

Requisitioned By ______________________ ______________________ Purchasing Agent

Shipped ______________________ Per ______________________

JOB COPY

FORM 8.42

Key Use of Form:

Order material.

Who Prepares:

Personnel in the purchasing department.

Who Uses:

The supplier for fulfillment. A copy of it is sent to the designated person who will be responsible for receiving the material. A copy will be retained by the purchasing department.

How to Complete:

Provide the information requested in the appropriate spaces.

Alternative Forms:

A company can revise the form to meet its specific needs.

8.43: Purchase Order Log Sheet

PURCHASE ORDER LOG SHEET

Period ______________________

P.O. Log	Vendor	P.O. No.	Purchase Order					Date Matl. Recd.	Other Chg. No.	Cost	Cumul. Total	Paid	Description/ Comments
			To Purch.	From Purch.	To Est. Dept.	Date Signed	Date Recd.						

FORM 8.43

Key Use of Form:

To keep track of the material ordering process.

Who Prepares:

Designated person in purchasing department.

Who Uses:

Purchasing personnel.

How to Complete:

Provide the information requested in the appropriate spaces.

Alternative Forms:

A company can revise the form to meet its specific needs.

8.44: Facilities Organization Material List

FACILITIES ORGANIZATION MATERIAL LIST

Listed By:	Project Manager	Job Description	Log No.	Bldg. Flr.
Designer:	Area		Mat'l. List No.	Sheet ____ of ____
Project Supervisor	Planner/Estimator		Date	Date Wanted

Item No.	Quantity	Lead Time	P.O. Number or Vendor	Description of Material	Remarks or Drawing No.

Copies to:

8.44: *(Continued)*

FORM 8.44

Key Use of Form:

To document all material requests to perform the job along with other pertinent information.

Who Prepares:

Personnel in the purchasing department.

Who Uses:

Plant engineering and purchasing personnel.

How to Complete:

Provide the information requested in the appropriate spaces.

Alternative Forms:

Form can be revised to meet specific company needs.

8.45: Requisition for Store Items

REQUISITION FOR STORE ITEMS

Department ______________________________ Date ______________

Person Completing Form ______________________________________

Item No.	Description	Quantity	Unit Cost	Total Cost

FORM 8.45

Key Use of Form:

Obtain material from in-house sources such as a storeroom or warehouse.

Who Prepares:

Designated person in the maintenance department.

Who Uses:

Person responsible for fulfilling the order.

How to Complete:

Provide the information requested in the appropriate spaces.

Alternative Forms:

Revisions can be made to make form company-specific.

8.46: Maintenance Equipment and Materials Inventory Checklist

MAINTENANCE EQUIPMENT AND MATERIALS INVENTORY CHECKLIST

Item No.	Description of Equipment and Materials	Amount Owned	No. on Hand	Checked By	Date

FORM 8.46

Key Use of Form:

To control the company's inventory of maintenance equipment and/or material.

Who Prepares:

Designated person responsible for maintaining inventory control.

Who Uses:

Designated department for review and filing for future use when material and equipment are required to perform maintenance work. A copy is retained by the organizational unit responsible for maintaining the inventory.

How to Complete:

The person taking the inventory will note the description of each item in the space provided along with the various quantities.

Alternative Forms:

Form can be revised to meet specific company needs.

8.47: Maintenance Tool Inventory Checklist

MAINTENANCE TOOL INVENTORY CHECKLIST

Item No.	Description of Part	No. Owned	No. on Hand	Checked By	Date

FORM 8.47

Key Use of Form:

To control the company's inventory of maintenance tools.

Who Prepares:

Designated person responsible for maintaining inventory control.

Who Uses:

Designated department for review and filing for future use when tools are required to perform maintenance work. A copy is retained by the organizational unit responsible for maintaining the inventory.

How to Complete:

The person taking the inventory will note the description of each item in the space provided along with the various quantities.

Alternative Forms:

Form can be revised to meet specific company needs.

8.48: Tool Sign-Out List

TOOL SIGN-OUT LIST

Date	Tool Description	Worker's Signature	Supervisor's Initials When Returned

FORM 8.48

Key Use of Form:

Document and control the checking out and returning of maintenance tools.

Who Prepares:

Designated person responsible for storing and maintaining tools.

Who Uses:

Same person who prepares.

How to Complete:

Provide the information requested in the appropriate spaces when tools are checked out and returned.

Alternative Forms:

A company can make appropriate revisions to this form to make it fit its needs.

J. CONTROL AND EVALUATION FORMS

Controlling the maintenance activity is monitoring it to see that it is being performed for the estimated costs and being accomplished in the scheduled time. It also entails evaluating whether the actual resources used were those that were identified in type and quantity in the job plan. In addition, many unplanned-for events can occur during the duration of the job and changes made to ensure the job is completed within the estimated time and budget.

There are two major categories of maintenance: preventive and crash. The first can be planned for and the second can only be anticipated. Because of this, it is not unusual to acquire a backlog of maintenance activities since management never knows exactly the magnitude of the resources needed to perform all the maintenance requested. To control backlog, the Maintenance Backlog Listing, Maintenance Work-Order Backlog and Performance Report, and Weekly Labor Backlog Report were provided in the previous section.

Another item related to backlog is overtime. This is required when a maintenance activity gets behind schedule and more time needs to be worked to complete it, or when a schedule is reduced in length for production reasons and the activity must be completed in a shorter amount of time. The Overtime Requirement form has been provided for use in ordering and controlling the use of overtime. Various forms have also been included relative to controlling the resource of labor. These include the Job Description Form, Supervisor's Personnel File, Employee Time Report, Attendance Record, and the Daily Absentee Report.

Forms used to help ensure a productive labor force are the Notice of Violation of Rules, Employee's Disciplinary Action, Daily Travel Diary and Memorandum, Inter-Office Communication, and New Maintenance Employee Training and Evaluation forms. The purpose of these forms is evident from their title and are self-explanatory in their completion. Once completed by the appropriate supervisor, the forms should be kept in a secure place for future use in evaluating the maintenance personnel along with the effectiveness of the entire maintenance organization.

Formal evaluation and analysis is a must if management is to ascertain if the work was carried out effectively and efficiently. This is accomplished utilizing the data contained on all the forms discussed above plus those of forms such as the Supervisor's Daily Checklist and Work Appraisal Form.

Whenever equipment fails, operations usually cease. In fact for this and other reasons downtime occurs while the equipment or other item is being repaired. One of the goals of management is to minimize downtime. To help them in this effort the Equipment Operator's Report, Equipment Failure and Accident Report, Breakdown Report, Maintenance Downtime Report, and Maintenance Failure History Analysis forms are provided.

8.49: Maintenance Backlog Listing

MAINTENANCE BACKLOG LISTING

Date ________________ Person Completing Form ________________________________

Craft ______________ No. of Workers ______________ Week Ending ____________

Total Hours Available Per Week ____________

Work Order Number	Date of Issue	Type of Maintenance	Equipment Designation	Location	Estimated Hours
			BACKLOG TOTAL		

FORM 8.49

Key Use of Form:

Used each week to list all the work that is backlogged or hasn't been assigned or completed.

Who Prepares:

Person responsible for tracking backlog (usually in scheduling department).

Who Uses:

The completed form is transmitted to the person responsible for scheduling so the work can get assigned on a timely basis and the backlog reduced. In most maintenance organizations, the planning and scheduling unit is responsible for identifying and controlling backlog.

How to Complete:

Provide the information requested in the appropriate spaces.

Alternative Forms:

Can make revisions to the form to make it company-specific.

8.50: Maintenance Work-Order Backlog and Performance Report

MAINTENANCE WORK-ORDER BACKLOG AND PERFORMANCE REPORT

Form Completed By ______________________________

Craft or Type of Work	Hrs./Wk. for W.O.s	Backlog as of (Date) ________ "Available" to Schedule			"Unavailable" to Schedule		
		Hours	Weeks	Avg. Time	Hours	Weeks	Avg. Time
TOTALS			X	X		X	X

Performance Last Month	
Percent Compliance	Percent Unscheduled
X	X

PERFORMANCE DEFINITIONS:

$$\text{Percent Compliance} = \frac{\text{No. of Scheduled Jobs Completed} \times 100}{\text{No. of Jobs Scheduled to Be Completed}}; \quad \text{Goal} \geqslant 90\%$$

$$\text{Percent Unscheduled} = \frac{\text{No. of Unscheduled Jobs Completed} \times 100}{\text{Total No. of Jobs Completed}}; \quad \text{Goal} \leqslant 10\%$$

8.50: *(Continued)*

FORM 8.50

Key Use of Form:

Used to keep track of backlog and also to evaluate the performance of the maintenance organization in reducing it.

Who Prepares:

Personnel in the planning and scheduling department.

Who Uses:

The evaluation will be performed by the designated person within the department and then reviewed by the appropriate manager who in turn takes any needed follow-up action to correct any deficiencies. The form itself is kept on file with copies going to supervisory staff involved with the backlog.

How to Complete:

Provide the information requested in the appropriate spaces.

Alternative Forms:

Revisions can be made to make this form company-specific.

8.51: Weekly Labor Backlog Report

WEEKLY LABOR BACKLOG REPORT

Prepared by ______________________ For Week Of ______________

	Total	Laborers	Mechanics	Electricians	Inst.	Carpenters	Painters
A. Total hourly workers in section							
B. Expected absences and							
scheduled vacations							
C. Assigned to nonscheduled work							
D. Worker-days available							
(A-B-C) × 5							
Authorized Work Backlog Available for Scheduling (Worker-Days)							
E. Work orders dated for this week							
F. Estimated new work orders							
dated for this week							
G. Total work dated for							
completion this week							
H. Work order not dated this							
week—can be scheduled							
I. Total authorized work avail-							
able for this week (G + H)							
J. Backlog—weeks (I ÷ D)							

FORM 8.51

Key Use of Form:

Used to keep track of backlog and also to evaluate the performance of the maintenance organization in reducing it.

Who Prepares:

Personnel in the planning and scheduling department.

Who Uses:

The evaluation will be performed by the designated person within the department and then reviewed by the appropriate manager who in turn takes any needed follow-up action to correct any deficiencies. The form itself is kept on file with copies going to supervisory staff involved with the backlog.

How to Complete:

Provide the information requested in the appropriate spaces.

Alternative Forms:

Revisions can be made to make this form company-specific.

8.52: Overtime Requirement

OVERTIME REQUIREMENT

Prepared By ____________________ Date __________

Date Needed	Unit or Name	Hours	Explanation or Reason

FORM 8.52

Key Use of Form:

To request overtime work.

Who Prepares:

Supervisor needing to have work performed (on an overtime basis).

Who Uses:

Designated personnel in scheduling department.

How to Complete:

Provide the information requested in the appropriate spaces.

Alternative Forms:

A company can make revisions to the form to meet its specific needs.

8.53: Job Description Form

JOB DESCRIPTION FORM

Job Title ______________________ Department ______________

Immediate Supervisor's Title ______________________________

Line of Promotion From ______________________ To ______________

Function of Job:

List or Description of Duties:

FORM 8.53

Key Use of Form:

To document the title, function, and duties entailed in each job classification. It is used when assigning work and evaluating the effectiveness of the person covered by the description.

Who Prepares:

The personnel responsible for maintaining job descriptions with input from all supervisory staff.

Who Uses:

It is maintained in central files with copies transmitted to designated maintenance managers to be used in the hiring, training, and evaluation of employees.

How to Complete:

Provide the information requested in the appropriate spaces.

Alternative Forms:

Revisions can be made to this form to make it company-specific.

8.54: Supervisor's Personnel File

SUPERVISOR'S PERSONNEL FILE

Date ____________________

Name ______________________________ Social Security No. ______________

Last address ______________________ Phone ______________________

____

Moved to ____

Moved to ____

Persons who will know where person is ______________________________

____

Occupation ____

Worked last on what job ______________________________________

In what capacity ____

Strong Skills ____

Can operate what equipment ______________________________________

Weaknesses ____

Last position in company ______________________ Salary ______________

Other comments ____

____

____

FORM 8.54

Key Use of Form:

To document important information about each employee.

Who Prepares:

Supervisor or the designated representative.

Who Uses:

Supervisor or other person who may need workers.

How to Complete:

Provide the information requested in the designated spaces.

Alternative Forms:

Revisions can be made to produce a customized form for a company.

8.55: Employee Time Report

EMPLOYEE TIME REPORT

Name ______________________________ Employee No. ______________

Week Ending ________________________

Work Order No.		M	T	W	T	F	S	S	Total
		Hours							
	STD	___	___	___	___	___	___	___	___
________	OT	___	___	___	___	___	___	___	___
	STD	___	___	___	___	___	___	___	___
________	OT	___	___	___	___	___	___	___	___
	STD	___	___	___	___	___	___	___	___
________	OT	___	___	___	___	___	___	___	___
	STD	___	___	___	___	___	___	___	___
________	OT	___	___	___	___	___	___	___	___
	STD	___	___	___	___	___	___	___	___
________	OT	___	___	___	___	___	___	___	___
TOTAL HOURS:		___	___	___	___	___	___	___	___

SUPERVISOR'S APPROVAL ______________________________

FORM 8.55

Key Use of Form:

To document amount of time spent in performing specific jobs for each day of the week.

Who Prepares:

The crew supervisor or the designated representative completes each week or pay period.

Who Uses:

Used as the basis of generating payroll checks by the accounting department and later to analyze the efficiency of the maintenance labor force by designated personnel in the plant engineering or maintenance department.

How to Complete:

Provide the information requested in the designated spaces.

Alternative Forms:

Revisions can be made to produce a customized form for a company. Also consider the use of Form 8.56.

8.56: Attendance Record

ATTENDANCE RECORD

Pay Period Ending ____________________

Employee Name ____

Department Number ____

Employee Signature ____

Date	Regular		Hours	Overtime to be Paid					Paid	Code	No Pay
	In	Out	Wrkd.	1/2	1	1 1/2	2	Total	Absence	*	Absence
16											
1 / 17											
2 / 18											
3 / 19											
4 / 20											
5 / 21											
6 / 22											
7 / 23											
8 / 24											
9 / 25											
10 / 26											
11 / 27											
12 / 28											
13 / 29											
14 / 30											
15 / 31											
Period Totals										X	

Total Overtime:		*Paid Absence Codes:	
Half	________	Illness	–I
Straight	________	Holiday	–H
Holiday-Comp.	________	Vacation	–V
Time & Half	________	Comp. Time	–C
Double	________	Jury Duty	–JD
		Death in Family	–D
Total	________	Other	–O

Shifttime Hours ________ Rate ________ = Total $ ________

Approved by ____

8.56: *(Continued)*

FORM 8.56

Key Use of Form:

To document amount of time spent in performing specific jobs for each day of the week.

Who Prepares:

The crew supervisor or the designated representative completes each week or pay period.

Who Uses:

Used as the basis of generating payroll checks by the accounting department and later to analyze the efficiency of the maintenance labor force by designated personnel in the plant engineering or maintenance department.

How to Complete:

Provide the information requested in the designated spaces.

Alternative Forms:

Revisions can be made to produce a customized form for a company. Also consider the use of Form 8.55.

8.57: Daily Absentee Report—Maintenance

DAILY ABSENTEE REPORT—MAINTENANCE

Date ____________________

Badge No.	Employee	Shift Crew No.	Supervisor	Reason for Absence	Replacement

FORM 8.57

Key Use of Form:

To document and control absenteeism.

Who Prepares:

Maintenance supervisor.

Who Uses:

Maintenance supervisor to determine and document reason for absenteeism and central management to evaluate the efficiency of maintenance personnel and supervision.

How to Complete:

Provide the information requested in the designated spaces.

Alternative Forms:

Revisions can be made to produce a customized form.

8.58: Notice of Violation of Rules

NOTICE OF VIOLATION OF RULES

Department ______________________________ Date ______________

To ____________________________

Payroll No. ________ Occupation ________ Service _____ yrs. _____ mos.

On ______________________ you were observed or found in violation of plant rule or regulation ____

Your actions were as follows: ______________________________________

____

____

____

Previous violations:

You are disciplined as follows:

☐ Reprimanded ☐ Suspended ☐ Discharged

____

____

____

Signed: ______________________________ (Supervisor)

FORM 8.58

Key Use of Form:

To formally notify an employee of violation of rules.

Who Prepares:

Immediate supervisor.

Who Uses:

Employer.

How to Complete:

Provide the information requested in the designated spaces.

Alternative Forms:

Revisions can be made to produce a customized form for a company. Form 8.59 may be used for same purpose.

8.59: Employee's Disciplinary Action

EMPLOYEE'S DISCIPLINARY ACTION

Name ______________________________ Date ______________

You are hereby given ______________________________ for the following infraction(s) of established rules:

Future infraction(s) will result in further disciplinary action and possible termination.

Supervisor's Signature ______________________________ Date ______________

Employee's Signature ______________________________ Date ______________

FORM 8.59

Key Use of Form:

To formally notify an employee of violation of rules.

Who Prepares:

Immediate supervisor.

Who Uses:

Employer.

How to Complete:

Provide the information requested in the designated spaces.

Alternative Forms:

Revisions can be made to produce a customized form for a company. Form 8.58 may be used for same purpose.

8.60: Daily Travel Diary and Memorandum

DAILY TRAVEL DIARY AND MEMORANDUM

Report No. ____________ Completed By ____________________

Sheet ______ of ______

Date	Day of Week	Place	Memorandum

FORM 8.60

Key Use of Form:

To record travel.

Who Prepares:

Person traveling.

Who Uses:

Supervisor of person traveling.

How to Complete:

Provide the information requested in the designated spaces.

Alternative Forms:

Revisions can be made to produce a customized form for a company.

8.61: Interoffice Communication

INTEROFFICE COMMUNICATION

TO	Office
From	Office
Subject	Date

Message

Signed

Reply

Date	Signed

8.62: New Maintenance Employee Training and Evaluation Schedule

NEW MAINTENANCE EMPLOYEE TRAINING AND EVALUATION SCHEDULE

Name ______________________________

Employee Number ____________________ Skill Type ____________________

Assignment			
Week	**Date**	**Shift**	**Evaluator/Trainer**
1			
2			
3			
4			
5			
6			
7			
8			
9			
10			

Note to Evaluators: At the end of each week, the trainer/evaluator must provide an evaluation of the employee.

FORM 8.62

Key Use of Form:

To document training and evaluate the person receiving the training.

Who Prepares:

Supervisor, trainer, or evaluator.

Who Uses:

Immediate and central maintenance management to review and evaluate a worker's training performance.

How to Complete:

Provide the information requested in the designated spaces.

Alternative Forms:

Revisions can be made to produce a customized form for a company.

8.63: Supervisor's Daily Checklist

SUPERVISOR'S DAILY CHECKLIST

Completed By ______________________________ Date ______________

Name	Attend-ance		Work Quality								Output		Attitude		Other		Notes
	Good	Poor															

FORM 8.63

Key Use of Form:

To evaluate the performance of each employee.

Who Prepares:

Crew supervisor.

Who Uses:

The completed form should be retained for incorporation into follow-up activities such as formal evaluation or termination. A copy should be sent to the central maintenance office for future reference.

How to Complete:

Space is available at the top of the form to place one's own rating criteria such as good, average, and poor. Every worker in the crew should be evaluated daily as to attendance, work quality, output, attitude, and other parameters. This is usually done on an annual or semi-annual basis by the person's supervisor. It is shared with the employee during the formal evaluation.

Alternative Forms:

Revisions can be made to produce a customized form for a company. Could also use Form 8.64.

8.64: Work Appraisal Form

WORK APPRAISAL FORM

☐ Annual ☐ Special ☐ Probationary

Name ______________________________

☐ 1st mo. ☐ 2nd mo. ☐ 3rd mo. Classification ______________

☐ 4th mo. ☐ Special ☐ Special Period from ________ to ________

I. Performance Appraisal (rate each item by selecting the phrase most closely describing the employee's actual work performance).

1. **Job Knowledge**
   a. Area work
   _____ Needs more training and experience
   _____ Able to perform most of the work
   _____ Knows enough to get by
   _____ Able to perform all work well
   _____ Able to perform most of the routine work
   _____ No experience in past year
   b. Floor work
   _____ Able to perform most of the routine work
   _____ Needs more training and experience
   _____ Able to perform all work well
   _____ Able to perform most of the work
   _____ Knows enough to get by
   _____ No experience in past year
   c. Public function setups
   Moving
   Window washing
   _____ Able to perform all work well
   _____ Able to perform most of the routine work
   _____ Able to perform most of the work
   _____ Needs more training and experience
   _____ Knows enough to get by
   _____ No experience in past year
2. **Ability to Improve**
   _____ Has mastered all duties of the position and still has the capacity to do more
   _____ Unable to do most of the routine work
   _____ Lacks ability to perform the work required by this position

8.64: *(Continued)*

_____ Has mastered most of the duties and still has the capacity to do more

_____ The routine work is the best that can be done and it causes trouble sometimes

3. **Quantity of Work**

_____ Has difficulty in organizing time in order to produce acceptable quantity of work. Has to be told what to do most of the time

_____ Self-starter—makes above-average use of work time

_____ Needs prodding or work time will be unproductive

_____ Makes average use of work time and achieves average quantity of work

_____ Self-starter—makes outstanding use of work time

_____ ______________________________

_____ ______________________________

4. **Quality of Work**

_____ Work is outstanding, rarely find items that need improvement

_____ Work is above average, rarely find examples that have been skipped or neglected

_____ Quality of daily and periodic work is average

_____ Weekly checks are necessary

_____ Daily checks are necessary

_____ ______________________________

_____ ______________________________

5. **Attitude Toward Work**

_____ Appreciates help and criticism

_____ Downgrades the company and many of the employees

_____ Does what is expected

_____ Respects the company and his/her position in organization

_____ Just another job—makes caustic remarks

_____ ______________________________

_____ ______________________________

6. **Initiative**

_____ Little evidence of initiative noticed during working hours

_____ Develops solutions to problems

_____ No indication of initiative during working hours

_____ Develops workable solutions to most problems on his/her own but keeps supervision informed

_____ Relies on others but follows suggestions

_____ ______________________________

8.64: *(Continued)*

7. **Loyalty**

_____ Critical of the company
_____ Willing to talk about the company in a positive way
_____ Inclined to be one-sided and negative about the company
_____ Self-starter, always willing to explain the company in a positive way
_____ If others insist, he/she will talk about the company in positive terms
_____ ____
_____ ____

8. **Personality**

_____ Talker but seems to get along with others
_____ Quiet but gets along with others
_____ Gets along with others
_____ Creates some feelings against self
_____ Respected by others
_____ ____

9. **Dependability**

_____ Average in work production and average attendance record
_____ Above average in work production with a perfect attendance record
_____ Average in work production with an unexplained attendance record average of one day sick every two months
_____ Barely average in work production with an above-average attendance record
_____ Above average in work production with an above-average attendance record
_____ ____
_____ ____

10. **Accident Prevention**

_____ Talks a lot about safety but doesn't do much about it
_____ Takes chances
_____ Works safely most of the time
_____ Sets a good example in safe work habits and encourages others to do the same
_____ Doesn't take chances but shows little interest in accident prevention
_____ ____
_____ ____

11. **Physical Makeup**

_____ Complains that work is too hard and the job is too large
_____ Accomplishes work with energy to spare
_____ Accomplishes work

8.64: *(Continued)*

____ Appears tired most of the time

____ Accomplishes work easily

____ ____

____ ____

12. **Cooperation**

____ Is argumentative

____ Is an outstanding team worker

____ Always does what is asked

____ Seldom voices objections

____ Always does a little more than is asked of him/her

____ ____

____ ____

II. General Comments

Comments on outside (second) job ____________________

Shift preference ____________________

Health ____________________

Attendance ____________________

Progress ____________________

Suggestions for self-improvement or development ____________________

____

Other comments ____________________

____

____________________ ____________________

(Date) Signature of person making work appraisal

III. Comments or suggestions by shift supervisor ____________________

____

____

____

IV. Comments or suggestions by other members of office supervision ____________________

____

____

____

This work appraisal has been discussed with me

____________________ ____________________

(Signature of employee) (Date)

____

(Signature of person explaining work appraisal)

8.64: *(Continued)*

ADDITIONAL COMMENTS:

Probationary ☐ Certified ☐ Unacceptable ☐

Below Average ☐ Average ☐ Above Average ☐ Outstanding ☐

COMMENTS

FORM 8.64

Key Use of Form:

To evaluate the performance of each employee.

Who Prepares:

Crew supervisor.

Who Uses:

The completed form should be retained for incorporation into follow-up activities such as formal evaluation or termination. A copy should be sent to the central maintenance office for future reference.

How to Complete:

Space is available at the top of the form to place one's own rating criteria such as good, average, and poor. Every worker in the crew should be evaluated daily as to attendance, work quality, output, attitude, and other parameters. This is usually done on an annual or semi-annual basis by the person's supervisor. It is shared with the employee during the formal evaluation.

Alternative Forms:

Revisions can be made to produce a customized form for a company. Could also use Form 8.63.

8.65: Equipment Operator's Report

EQUIPMENT OPERATOR'S REPORT

Person Completing Form ______________________ Date __________

Equipment ______________________ Location ______________________

Time Report Completed ______________________

Description of Problem (be specific):

FORM 8.65

Key Use of Form:

To repair an equipment problem.

Who Prepares:

Equipment operator.

Who Uses:

The form is sent to the equipment operator's supervisor who in turn initiates a work request to correct the problem. A copy is retained by the supervisor until the work has been performed.

How to Complete:

Provide the information requested on the form in the designated spaces.

Alternative Forms:

A company can revise the form to meet its specific needs.

8.66: Equipment Failure and Accident Report

EQUIPMENT FAILURE AND ACCIDENT REPORT

Location ____________________ Date of Report ____________

Completed By ____________________ Date of Occurrence ____________

Equipment, System, or Building Identification
Describe Failure and Extent of Damage and Apparent Reason for Failure
Estimated Cost to Repair and Restore
Estimated Date of Completion
How is Future Occurrence Prevented?
Suggested Design and Specification Changes
Other Comments and Recommendations

FORM 8.66

Key Use of Form:

To report equipment failure and/or accident.

Who Prepares:

Crew or department supervisor.

Who Uses:

Designated management personnel to initiate corrective action to repair the equipment.

How to Complete:

Provide information requested in the appropriate spaces.

Alternative Forms:

Company may revise this form to meet its specific needs.

8.67: Breakdown Report

BREAKDOWN REPORT

A. WHAT MACHINE BROKE DOWN? DATE: ______

Department ______

Line ______ ASSET NO.: ______

Machine ______

B. WHAT BROKE OR FAILED, AND WHAT CAUSED IT TO HAPPEN?

______

______

______

C. WHAT WAS DONE TO REPAIR FAILURE AND RESTORE OPERATION?

______

______

______

D. HOW MANY MINUTES OF LINE PRODUCTION TIME WERE LOST?

______ (@ ______ \$/min = \$ ______)
↑——by planner——↑

E. HOW MANY TOTAL WORK HOURS WERE USED FOR PART C? ______

F. WHAT CAN BE DONE TO PREVENT A SIMILAR FAILURE FROM OCCURRING IN THE FUTURE?

1. Would preventive maintenance inspections have prevented? Yes ____ No ____
2. Should a breakdown analysis be initiated? Yes ____ No ____
3. Can corrective action be taken to prevent recurrence? Yes ____ No ____
4. If yes to No. 3, what action do you recommend, and can it be done now?

______

______

______

______
Maintenance Supervisor

G. ACTION TO BE TAKEN BY MAINTENANCE PLANNING SECTION:

1. Perform breakdown analysis ☐
2. Initiate action as recommended in F. 4, above ☐
3. Take no further action ☐
4. Other: ______

______
Maintenance Engineer

8.67: *(Continued)*

FORM 8.67

Key Use of Form:

To report a breakdown of equipment.

Who Prepares:

Maintenance supervisor.

Who Uses:

Designated plant engineering or maintenance department personnel who will schedule and order follow-up work to repair equipment.

How to Complete:

Provide the information requested in the appropriate spaces.

Alternative Forms:

Company may revise this form to meet its specific needs.

8.68: Maintenance Downtime Report

MAINTENANCE DOWNTIME REPORT

For Lost Time of 10 Minutes or More Per Shift

Equipment ____________________ Date ____________ Line ____________

Time Lost ____________________

Reason ____

____

____

____

____

I Have Taken the Following Action to Correct This Problem ____________________

____

____

____

____

____

____

____

Maintenance Supervisor Originating ________________________________

Other Maint. Supervisor Initial & Date Here ________________________________

FORM 8.68

Key Use of Form:

To document lost time of 10 minutes or more.

Who Prepares:

Immediate maintenance supervisor.

Who Uses:

Person responsible within plant engineering or maintenance department who evaluates performance, plans schedules, and orders work to alleviate the problem.

How to Complete:

Provide the information requested in the appropriate spaces.

Alternative Forms:

Company may revise this form to meet its specific needs.

8.69: Maintenance Failure History Analysis

MAINTENANCE FAILURE HISTORY ANALYSIS

Date ________ Week Ending ________ Person Completing Form ________

Craft ________

Equipment Designation	Location	Maintenance Activity	Average Time Between Failures	Last Performance	Labor Costs	Material Costs	Total Costs

TOTAL

FORM 8.69

Key Use of Form:

To document and evaluate reason for equipment failure.

Who Prepares:

Department supervisor.

Who Uses:

Central maintenance or plant engineering personnel .

How to Complete:

Provide the information requested in the appropriate spaces.

Alternative Forms:

A company can revise the form to meet its specific needs.

8.70: Service/Maintenance Contract Authorization

SERVICE/MAINTENANCE CONTRACT AUTHORIZATION

Date ____________ Authorization No. ____________

Dept. Name ____________ Dept. No. ____________

Requestor ____________ New ______ Renewal Request ______

Describe equipment being serviced (type and quantity) ____________

____

____

Describe service/maintenance to be provided ____________

____

____

Service History ____________

____

____

Contractor ____________ ESTIMATED COST

____________ ____________

Contract Period ____________ to ____________

APPROVALS	SIGNATURE	DATE

FORM 8.70

Key Use of Form:

To document authorization for a service or maintenance contract with an external party.

Who Prepares:

The person responsible for contract administration within the company.

Who Uses:

Central maintenance or plant engineering personnel .

How to Complete:

Provide the information requested in the appropriate spaces.

Alternative Forms:

A company can revise the form to meet its specific needs.

8.71: Weekly Summary Outside Contractor Report

WEEKLY SUMMARY OUTSIDE CONTRACTOR REPORT

Date ______________________

Completed By ______________________ Week Beginning ______________________

Date	Job No.	Description	Contractor	Hours

FORM 8.71

Key Use of Form:

Summarize labor hours expended in performing work by outside contractor.

Who Prepares:

Designated person in the company responsible for working with and managing the work of outside contractors.

Who Uses:

Central management to evaluate efficiency of outside contractors.

How to Complete:

Provide the information requested in the appropriate spaces.

Alternative Forms:

A company can revise the form to meet its specific needs.

K. BUILDING ENERGY USE FORMS

The last series of forms included in this section are to be used in evaluating and controlling the use of energy to light, heat, and cool the building(s). They are the Monthly Plant Energy Consumption, Energy Conservation Project Evaluation Summary, Energy-Saving Survey, and Energy Conservation Capital Projects forms. To complete these forms, provide the information requested in the appropriate spaces. The personnel responsible for completing these forms will depend on the type, size, and organization of the company. For smaller companies, it may be a person in the maintenance department; for larger companies, plant engineering or even an outside contractor. The completed forms will be retained by the central organizational unit for use in evaluating energy consumption.

8.72: Monthly Plant Energy Consumption

MONTHLY PLANT ENERGY CONSUMPTION

Form Completed By ______________________ Date ____________

	Electric Power			Natural Gas			Fuel Oil			Coal			TOTAL Btu	Number of Units Produced	Btu Per Unit of Production
Year	kWh	Btu/kWh	Btu	k cu ft	Btu/k cu ft	Btu	gal	Btu/gal	Btu	TONS	Btu/lb	Btu			
Jan.															
Feb.															
Mar.															
Apr.															
May															
June															
July															
Aug.															
Sep.															
Oct.															
Nov.															
Dec.															
Year															
Jan.															
Feb.															
Mar.															
Apr.															
May															
June															
July															
Aug.															
Sep.															
Oct.															
Nov.															
Dec.															

8.73: Energy Conservation Project Evaluation Summary

ENERGY CONSERVATION PROJECT EVALUATION SUMMARY

Capital ____________________ or Expense ____________________

Department ____

Date ____

Project No. ______________ Person Responsible ______________________________

Project Title ____

Description of Project ____

____

____

____

____

Location ____

____

Financial Evaluation

Estimated

Energy saving (electric power kWh/yr, steam lb/yr, etc.)

Utility or Raw Material	Saving
______________________	______________________ /yr
______________________	______________________ /yr
______________________	______________________ /yr
Total energy saving	______________________ MBtu/yr
Total energy cost saving	______________________ $/yr
Other cost saving due to:	
______________________	______________________ $/yr
Additional cost due to:	
______________________	______________________ $/yr
Net cost saving	______________________ $/yr
Cost of project	______________________ $

8.74: Energy-Saving Survey

ENERGY-SAVING SURVEY

Department ______________ Surveyed By ______________ Date of Survey ______________

Fuel Gas or Oil Leaks	Steam Leaks	Compressed Air Leaks	Condensate Leaks	Water Leaks	Damaged or Lacking Insulation	Excess Lighting	Excess Utility Usage	Equipment Running & Not Needed	Burners Out of Adjustment	Leaks of or Excess of HVAC	Location	Date Corrected

8.75: Energy Conservation Capital Projects

ENERGY CONSERVATION CAPITAL PROJECTS

Department ______________ Form Completed By ______________ Date ______________

Project Number	Project Description	Energy Savings Btu/Year	Capital Cost $	Ratio $\frac{\text{Btu/Year Savings}}{\text{\$ Capital}}$	Percent ROI	Priority	Status

Index

A

Absentee, Daily Report, 414
Aeration Equipment Inspection Checklist, 261
Air Compressor Unit Inspection Checklist, 260
Air Conditioning System Checklist, 205
Air Conditioning Unit Inspection Checklist, 259
Air Conditioning, Window and Fan Coil Unit Inspection Checklist, 258
Air-Cooled Condenser Inspection Checklist, 257
Air Filter Inspection Checklist, 256
Air Handling Unit Inspection Checklist, 244–245
Antenna Inspection Checklist, 21–22
Asphalt Surface Record, 99
Attendance Record, 412–413

B

Bakery Equipment Inspection Checklist, 224
Battery Inspection Checklist, 310
Bill of Material, 392
Breakwater Inspection Checklist, 107
Bridge and Trestle Inspection Checklist, 23–25
Building
  Cleaning Schedule, 134–135
  Exterior Survey Form, 68–69
  Finishes Inventory Form, 35–36
  Herbicide Log, 130
  Insecticide Log, 172
  Inspection Checklist, 76–77
  Interior Survey Form, 70–73
  Pesticide Log, 171
  Rating Profile Summary Sheet, 63–67
Building/Grounds
  Work Order, 119–120
  Work Schedule, 121
Budget Control Sheet, 379
Bulkhead Inspection Checklist, 111

C

Carpet and Upholstery Work Report, 142
Cathodic Protection System Inspection Checklist, 105
Chemical Feed Equipment Fresh Water Supply Inspection Checklist, 239–240
Chimney and Stack Inspection Checklist, 26–27
Chlorinator and Hypochlorinator Inspection Checklist, 238
Circuit Breaker, Oil, Inspection Checklist, 318
Cleaning, Daily Report, 139
Cooling Tower Inspection Checklist, 220
Custodial
  Area Assignment, 136–137
  Area Sanitation Report, 145–146
  Inspection Report, Floor Maintenance, 147–148
  Inspection Report, Window Maintenance, 149–150
Custodial and Maintenance Program, Evaluation Form, 156–161

D

Dehumidification Unit Inspection Checklist, 255
Disconnecting Switch Inspection Checklist, 307–308
Dishwashing Equipment and Accessories Inspection Checklist, 253–254
Doors, Power-operated Inspection Checklist, 314
Dust Control System Check Sheet, 262–263

E

Earthquake Valve and Pit Inspection Checklist, 217
Electrical
  Controls Service Record, 279–280
  Equipment Inventory Record, 275
  Ground Inspection Checklist, 293
  Pothead Inspection Checklist, 315
  Power Plant Inspection Checklist, 316
  Power Shutdown Notice, 327
Electrical Systems and Equipment, 273
Electrical Systems, Building, Inspection Checklist, 282
Electrical Systems, Waterfront, Inspection Checklist, 317
Electronic Air Cleaner Inspection Checklist, 309
Elevated Tank Inspection Checklist, 28
Elevator
  Call Checklist, 331
  Inspection Checklist, 300–301
Employee
  Disciplinary Action, 416
  Time Report, 411
Energy
  Energy Conservation Capital Projects, 438
  Energy Conservation Project Evaluation Summary, 436
  Energy-Saving Survey, 437
  Monthly Plant Energy Consumption, 435
Environmental Control System Pretesting and Startup Checklist, 187
Equipment
  Breakdown Report, 428–429
  Data Card, 178–179
  Failure and Accident Report, 427

Equipment *cont'd*
- History Form, 339
- Maintenance Record, 338
- Operator's Report, 426
- Record Form, 177
- Survey, 190–192

Equipment and Repair Record, 276
Estimating Formats
- Equipment Maintenance Cost Record, 380–381
- Estimate Breakdown, 377–378
- Maintenance Cost Data Sheet, 374–375
- Recapitulation Sheet, 376
- Summary Sheet, 372–373
- Work Sheet, 370–371

Evaluation Form 3, 403
Expediting Form, 391
Exterior Ground Facility Inventory, 97–98
Exterior Maintenance Quality Assurance Evaluation, 162–163
Eye Wash Fountain Inspection Checklist, 241

F

Facilities Organization Material List, 397–398
Facilities Project Work Report, 138
Fence Inspection Checklist, 110
Filter, Liquid, Inspection Checklist, 246
Finishes
- Exterior/Interior, 37
- Exterior/Interior Inspection Checklist, 60–62

Fire Alarm
- Box Inspection Checklist, 322
- Box Light Inspection Checklist, 321
- Panel Inspection Checklist, 320

Fire Protection Equipment, Annual Inspection Checklist, 203–204
Floor Maintenance, Daily Report, 140
Food Warmer/Grill Inspection Checklist, 323
Freezer, Walk-in Inspection Checklist, 231
Fuel
- Disconnect Inspection Checklist, 324
- Distribution Facility Inspection Checklist, 226–227
- Receiving Facility Inspection Checklist, 228
- Storage Inspection Checklist, 229–230

G

Garbage Disposal Inspection Checklist, 117
Gas
- Distribution System Inspection Checklist, 212
- Heating Inspection Checklist, 213–214

Generator, Emergency, Inspection Checklist, 312
Generator Service Record, 279–282
Glass Replacement Data, 58
Grasses, Trees and Shrubbery Inventory Form, 95–96
Grounds, 93
Grounds Inspection Checklist, 101–103

H

Heat Exchange Data Sheet, 181
Heater, Unit Inspection Checklist, 218
Heater and Console Control Inspection Checklist, 251
Heating and Ventilating System, Annual Inspection Checklist, 193–199
Heating and Ventilating Unit Inspection Checklist, 250
Heating, Ventilation and Air Conditioning Drawing Checklist, 183–186
Housekeeping, 131
- Inspection Form, 151–153
- Supervisor Checklist and Report, School Buildings, 154–155
- Supervisor's Daily Personnel Checklist, 173

I

Ice Maker Inspection Checklist, 249
Incinerator Inspection Checklist, 258
Inspection Forms
- Electrical Systems and Equipment, 281
- Finishes, Interior/Exterior, 59
- Housekeeping, 143
- Landscaping and Grounds, 100
- Maintenance Management, 355
- Mechanical Systems and Equipment, 182
- Structural Systems, 9

Inspection Record Sheet, 362
Interior and Exterior Finishes, 33
Interior and Exterior Inspection Checklist, 60–62
Interior Maintenance Quality Assurance Evaluation, 164–165
Inter-office Communications, 418
Inventory Forms
- Electrical Systems and Equipment, 274
- Finishes, Interior and Exterior, 34
- Landscaping and Grounds, 94
- Maintenance Management 334, 337
- Mechanical Systems and Equipment, 176
- Structural Systems, 6–8

J-K

Job Description Form, 409

L

Labor Backlog, Weekly Report, 407
Landing Ramp Inspection Checklist, 31
Landscaping and Grounds, 93
Laundry Equipment Inspection Checklist, 247
Light
- Navigation Inspection Checklist, 319
- Service Requisition, 330
- Tube Replacement Form, 329

Lighting, Emergency Inspection Checklist, 316
Lighting System Inspection Checklist, 283–284

Lightning Arrestor Inspection Checklist, 311
Liquid Filter Inspection Checklist, 246
Lubrication Schedule, 269–270

M

Maintenance
Backlog Listing, 404
Contract Organization, 432
Downtime Report, 430
Employee Training and Evaluation Schedule, 419
Equipment and Material Inventory Checklist, 400
Failure History Analysis, 431
Management, 333
Management Control Form, 403
Management Flow Chart 2
Material Requisition, 393
Preventive, Inspection, 359
Preventive Inspection Monthly Report, 361
Report, 358
Specification, 335–336
Survey Form, 356
Tool Inventory Checklist, 401
Weekly Summary Outside Contractor Report, 433
Material Order Form, 394
Mechanical Downtime Report, 267
Mechanical Service Overtime Report, 268
Mechanical Systems and Equipment, 175
Motor
Inspection Checklist, 294–295
Job Sheet, 328
Maintenance Record, 277–278
Motor and Generator Inspection Checklist, 296
Motor and Pump Inspection Checklist, 297
Motor and Pump Assembly (Hydraulic) Inspection Checklist, 298–299
Motor, Generator and Controls Service Record, 279–282

N

Notice of Violation of Rules, 415

O

Oil Consumption Form, 271
Overhead Crane Inspection Form, 32
Overtime Requirement, 408

P

Paint Spray Booth Inspection Checklist, 215
Painting
Building Work Record, 57
Historical Record, 56
Panel Board Inspection Checklist, 302
Pavement Inspection Checklist, 104
Pier Inspection Checklist, 111
Piping System Inspection Checklist, 208–209
Planning Formats 3, 363
Comprehensive Maintenance Planning Sheet, 367–368
Maintenance Department Planner, 364
Maintenance Planning Sheet, 365–366
Plant Engineering and Maintenance Survey Report, 357
Plumbing, Annual Inspection Checklist, 200–202
Pneumatic Temperature Controls Maintenance Checklist, 264–265
Pole
Steel Power, Inspection Checklist, 29
Wood, Inspection Checklist, 30
Pump Data Sheet, 180
Pump, Vacuum Producer, Inspection Checklist, 252
Pumping Plant, Potable Water, Inspection Checklist, 225
Purchase Order, 395
Purchase Order Log Sheet, 396
Purchasing Forms, 391

Q

Quaywall Inspection Checklist, 111

R

Railroad Crossing Signal Inspection Checklist, 114
Railroad Trackage Inspection Checklist, 112–113
Refrigeration Equipment Inspection Checklist, 234
Refuse/Garbage Inspection Checklist, 117
Reinforced Concrete Inspection Checklist, 10–16
Requisition, 39
Store Items, 399
Retaining Wall Inspection Checklist, 106
Room Housekeeping Checklist, 144
Roofs
Condition Survey Report, 82
Historical Record, 39
Asphalt Roll Roofing, 42–43
Asphalt Shingle, 54–55
Built-Up, 40–41
Cement Composition, 44–45
Metal, 46–47
Slate, 48–49
Tile, 50–51
Wood Shingle, 52–53
Inspection Checklist, 74–75
Asphalt Roll Roofing, 92
Asphalt Shingle Roof, 86
Built-Up, 84–85
Cement Composition, 87
Metal, 88–89
Slate, 90
Tile, 91
Installation Summary Specification, 37–38
Survey of Roof Condition, 78–81

S

Safety Shower Inspection Checklist, 235
Schedule Forms 3,, 382
  Building and Grounds, 121
  Building Cleaning, 134–135
  Daily Schedule, 388
  Housekeeping Master Schedule, 133
  Master Schedule, 385
  Monthly Master Maintenance Log, 386
  Office Facility Schedule Maintenance Routine, 389
  Project Master Schedule, 383–384
  Schedule Control Sheet, 390
  Vehicle Maintenance, 130
  Weekly Maintenance Work Schedule, 387
Seawall Inspection Checklist, 107
Septic Tank Inspection Checklist, 216
Sewage Collection and Disposal System Inspection Checklist, 232–233
Shop Area Quality Assurance Evaluation, 166–168
Shrubbery Inventory Form, 95–96
Snow Removal Call-In Sheet, 170
Stack Inspection Checklist, 26–27
Steam Trap Inspection Checklist, 219
Storm Drainage System Inspection Checklist, 115–116
Structural Steel Inspection Checklist, 17
Structural System, 5
  Inspection Form, 9
  Inventory Form, 6
Structural Wood Inspection Checklist,, 18–19
Supervisor
  Daily Checklist, 420
  Personnel File, 410
Swimming Pool
  Equipment Inspection Checklist, 236
  Inspection Checklist, 237
Switch Gear Inspection Checklist, 304

T

Telephone Line (Open Wire) Inspection Checklist, 305
Time Clock Inspection Checklist, 306
Tool Sign-Up List, 402
Transformer
  Distribution Checklist, 285–287
  Power, Inspection Checklist, 288–292
Travel, Daily Diary and Memorandum, 417
Tree Inventory Form, 95–96
Trestle Checklist, 23–25
Truss Inspection Checklist, 20
Turbine
  Large, Inspection Checklist, 222–223
  Small, Inspection Checklist, 221
Tunnel Inspection Checklist, 108–109

U

Underground
  Cable, Inspection Checklist, 325
  Structure, Inspection Checklist, 108–109

V

Vault and Manhole (Electrical) Inspection Checklist, 303
Vehicle
  Inspection Report, 124–125
  Maintenance Report, 122–123
  Maintenance Schedule, 128–129
  Preventive Maintenance Record, 126–127
Ventilating Fan and Exhaust System Inspection Checklist, 243

W

Wall Inspection Checklist, 110
Water Heater Inspection Checklist, 210–211
Water Storage, Fresh, Inspection Checklist, 242
Wharf Inspection Checklist, 111
Window Daily Washing Report, 141
Work Appraisal Form, 421–425
Work Order Formats
  Building and Grounds, 119
  Immediate Maintenance Request, 341
  Maintenance Service Request, 343
  Maintenance Work Order, 345–346
  Maintenance Work Request, 347–348
  Minor Maintenance Work Order, 344
  Repair Order, 342
  Service Order, 342
Work Order Summary Formats, 350
  Daily Service Log, 354
  Job Order Register, 352
  Labor and Material Work Order Summary, 353
  Maintenance Work Sheet, 351

X-Y-Z